THE OXFORD HISTORY OF
ENGLISH LITERATURE

General Editors
JOHN BUXTON, NORMAN DAVIS,
BONAMY DOBRÉE, *and* F. P. WILSON

III

THE OXFORD HISTORY OF ENGLISH LITERATURE

Certain volumes originally appeared under different titles (see title-page versos).
Their original volume-numbers are given below.

MALORY AND FIFTEENTH-CENTURY DRAMA, LYRICS, AND BALLADS

E. K. CHAMBERS

CLARENDON PRESS · OXFORD

Oxford University Press, Walton Street, Oxford OX2 6DP

Oxford New York Toronto
Delhi Bombay Calcutta Madras Karachi
Petaling Jaya Singapore Hong Kong Tokyo
Nairobi Dar es Salaam Cape Town
Melbourne Auckland

and associated companies in
Berlin Ibadan

Oxford is a trade mark of Oxford University Press

Published in the United States by
Oxford University Press, New York

First published 1945
Reprinted (with corrections) 1947, 1948, 1954, 1990

Originally published as volume II part 2 with the title English Literature at the Close of the Middle Ages
(ISBN 0-19-812203-9)

British Library Cataloguing in Publication Data

Chambers, E. K. (Edmund Kerchever)
[English literature at the close of the middle ages]
Malory and fifteenth-century drama, lyrics and ballads.—
(The Oxford history of English literature v. 3)
1. English literature 1400-1625. Critical studies
I. [English literature at the close of the middle ages]
I. Title
820.9'002
ISBN 0-19-812230-6

Library of Congress Cataloging in Publication Data

Chambers, E. K. (Edmund Kerchever), 1866-1954.
[English literature at the close of the Middle Ages]
Malory and fifteenth-century drama, lyrics, and ballads / E. K. Chambers.
p. cm.—(The Oxford history of English literature; 3)
Originally published: English literature at the close of the
Middle Ages. 1947. (The Oxford history of English literature; 2, pt. 2)
Includes bibliographical references.
English literature—Middle English, 1100-1500—History and
criticism. 2. Malory, Thomas, Sir, 15th cent. Morte d'Arthur.
3. Arthurian romances—History and criticism.   I. Title.
II. Series.
PR291.C5 1990   820.9'002—dc20   89-25493
ISBN 0-19-812230-6

Printed in Great Britain by
Courier International
Tiptree, Essex

EDITORS' NOTE

THIS half-volume, together with the half-volume on *Chaucer and the Fifteenth Century*, give a full treatment of English literature in the fifteenth century. For two reasons, however, Sir Edmund Chambers's book departs from the usual principle in this series of allotting a volume or half-volume to a particular period. The one is that the story of the medieval drama and the story of the ballad are better told without strict attention to the boundaries of time; and the other is that the editors have wished to give the author the opportunity of writing once more on subjects which have interested him in the past—the medieval drama, the carol and fifteenth-century lyric, and Malory. While the contents are more miscellaneous than elsewhere in the series, he has been able to give to the book the not inappropriate title, *The Close of the Middle Ages*.

CONTENTS

MEDIEVAL DRAMA

MEDIEVAL drama, in England as elsewhere, owes nothing to the tragedy and comedy of insolent Greece and haughty Rome. Before the Christian era began these were already a closed account. The plays of Seneca were probably intended for readers only. In the theatre of Pompey legendary stories were sung to the dancing of a *pantomimus*, and *mimi*, long branded with infamy, performed satirical and shameless farces. They did not spare the new religion, and in return the theatrical performances became the subject of many condemnations by ecclesiastical writers from the *De Spectaculis* of Tertullian onwards, and in more formal pronouncements by early councils of the western Church. If Augustine and others still retain some interest in the classical playwrights, it is as literature only, not as living drama. The degenerate theatre finally disappeared during the barbaric invasions of the sixth century, and the dispossessed *histriones*, as they were then called, were driven afoot, to merge with the descendants of the story-telling Teutonic poets, in the miscellaneous body of entertainers who haunted the towns and thoroughfares of the Middle Ages. They are still *mimi* and *histriones*, but also *ioculatores*, and, in so far as they became domesticated in courts or great houses, *ministeriales*, minstrels. Some tradition of impersonation, which, at any rate when accompanied with dialogue, amounts to drama, survived amongst them. There is playing of 'japis' or jests, as well as of more serious things, in the fourteenth century. 'L'uns fet l'ivre, l'autre le sot.' There are *estrifs* or disputes, and *débats*. As a body the minstrels inherit the clerical hostility evoked by the infamous element in their ancestry. They are condemned by the canon law, as codified by Gratian in the twelfth century, and by the *Decretals* of Gregory IX in the thirteenth. The *Penitential* of Thomas de Cabham, a little later, is more discriminating, and recognizes a higher as well as the lower element in minstrelsy. A few writers, in the twelfth century, and again in the fourteenth, use language which shows a consciousness of a theatrical origin for the lower element. Thus Walter Reynolds, who became Archbishop of Canterbury in 1314, is called a *mimus*, and said to have won the royal favour through his skill *in ludis theatralibus*. But it is the decadent Roman stage which such writers have in mind. Terence, it is true, continued to be studied, as a master of life, throughout the Middle Ages. In the tenth century Hrotsvitha, a nun of Saxony, took him as a model for half a dozen plays in glorification of chastity and the constancy of martyrs. It is unlikely that they were intended for representation. Possibly a true conception of the nature of classical drama was in the mind of the exceptionally learned John of Salisbury, who accompanies a typical twelfth-century denunciation of the *tota ioculatorum scena* by a definition of comedy as 'vita hominis super terram, ubi quisque sui oblitus personam exprimit alienam'. Imitation, of course, as here defined, is a fundamental instinct of humanity. It shows itself already in the seasonal *ludi* of the folk, who call the leaders of their revels kings and queens. But it is rare to find in a medieval writer any consciousness of an analogy between ·classical drama and the performances familiar to him ·in his own. day. Very occasionally, in the thirteenth and fourteenth centuries, a play is called a comedy. But normally this term and that of tragedy stand, as they stood for Dante, for varieties of narrative poetry, with a cheerful and a melancholy note respectively. Nor is the term 'theatre' normally applied to a medieval stage. It first appears at Exeter in 1348. 'Drama' is not an English word until the sixteenth century.

The human impulse to *mimesis* made a fresh start, where perhaps it might have been least expected, in the midst of the ecclesiastical liturgy itself. This, in the western Church, had reached its final development by the beginning of the tenth century, with the sacrificial Mass as its central feature, and the Office of.readings and praise and prayer, which, in the greater churches and monasteries at least, extended over the eight hours of the day from Matins to Compline. There was much singing, often divided between the priest and a choir, or between two parts of the choir, in what was called 'antiphon'. This naturally tended to approximate to the form of dialogue. The ritual was especially elaborate on the feast-days, which in their succession from the Annunciation to the Ascension commemorated the main incidents in the life of the Founder of Christianity. On these occasions it often included symbolical ceremonies, some of which are already traceable in the Church of Jerusalem during the fourth century, although others are first recorded later than the emergence of drama itself. At the Annunciation the deacon might hold a palm-branch while he read the Gospel, or a figure representing Gabriel might be lowered from the roof. At Christmas a *praesepe*, or crib, might be used as an altar, perhaps with a Mary and Joseph at its side. St. Francis, at Greccio in 1223, introduced a live ox and ass. At the Epiphany a mechanical star was suspended from the roof. At the Purification, which was also the Presentation of Christ in the Temple, might be carried into the church a lighted candle to symbolize the Light of the World, or a gospel-book, or an image of the Son and His Mother. On Palm Sunday came a procession with palm-branches to celebrate the entry into Jerusalem, sometimes with a gospel-book or a litter to represent the Saviour, or in Germany a wooden figure of Christ on an ass, known as a *Palmesel*. During Holy Week, when the gospel narratives of the Passion were sung, a curtain was often torn to represent the rending of the veil of the Temple, and at the *Tenebrae* of Matins and Lauds the altar candles were solemnly extinguished, one by one. On Holy or Maundy Thursday the feet of the poor were washed in church, in remembrance of the Last Supper. At the Ascension a cross or image was raised aloft, or a priest climbed the pulpit stairs. At Pentecost a censer and a mechanical dove came down from the roof. Such ceremonies at least hover upon the verge of drama, but they cannot be regarded as true drama, because they lack the essential element of impersonation, in which the actions and words are not those of the performer himself but of another whom he represents. That he usually also assumes the outward appearance and dress proper to that other is perhaps less essential. Impersonation is generally accompanied in drama by dialogue, but opinions differ as to whether that is a necessary feature. Perhaps the early ceremony which approximates most nearly to drama, as here defined, is one which does not belong to the festal calendar at all but was used in the ninth century at the dedication of a church. There is a dialogue between the bishop without and a clerk within *quasi latens*. The bishop demands that the King of Glory shall enter. Then he who is within *quasi fugiens egrediatur*. It is difficult not to feel that he is impersonating an evil spirit who is being expelled.

Some of these liturgical practices were taken up into the

new drama as it developed. But for its main origin we must look not to them but to the musical elaboration of the Mass itself; and here, as I have written elsewhere, we can trace it back, even 'beyond the very borders of articulate speech'. Church music had its full share in the literary and artistic movement of the ninth century, which is known as the Carolingian Renaissance. The traditional melodies were lengthened, in the first instance, by prolonged repetition of the final vowel sound of the *Alleluia*, which was sung as part of the Mass. Later the same practice was extended to other parts both of this and of the Office. Ultimately words were fitted to these melodies, some of which came to take stanzaic form. Technically these elaborations are called sequences, proses, or tropes. One of them, perhaps written by Tutilo of St. Gall in the tenth century, was originally prefixed to the introit of the Easter Mass. It went as follows:

Interrogatio:
Quem quaeritis in sepulchro, Christicolae?
Responsio:
Iesum Nazarenum crucifixum, o caelicolae.

Non est hic, surrexit sicut praedixerat; ite, nuntiate quia surrexit de sepulchro.

Later this was transferred to a procession between Tierce and Mass and thence to Matins, where it stood just before the final *Te Deum*. And here it became related to two symbolical ceremonies which I have not recorded above. Normally Mass was celebrated with a bread or Host consecrated during the service. On Good Friday alone there was no consecration, and the Host used was one which had been consecrated on the day before, and then laid aside in a receptacle which came to be known as a 'sepulchre'. This practice apparently served as a model for a more elaborate one on Good Friday itself. It attached itself to the ceremony, already known at Jerusalem in the fourth century, of the Adoration of the Cross, which took place, with psalms and antiphons and a kissing of the Cross, immediately before the introduction of the 'pre-sanctified' Host. By the tenth century at latest this was followed, either during the Mass itself or after Vespers, by a procession in which the Cross, often accompanied by a Host, was removed to a ' sepulchre' in the choir, with a ceremony known as the Deposition of the Cross.

This naturally involved an analogous Elevation of the Cross at dawn on Easter Day, in which the Cross was restored to its normal position in the choir. Occasionally the texts of the Deposition and Elevation include a hint of the *Descensus Christi ad Inferos*, as related in the apocryphal *Gospel of Nicodemus*. Finally, to the Deposition and Elevation was appended a third ceremony, the Visitation of the Cross, which took place on Easter Monday at Matins. And of this the germ appears to have been none other than the *Quem quaeritis* trope already described. But it has now become clearly a drama, for the essential element of impersonation has entered into it. As we find it described, not perhaps in quite its earliest form, by St. Ethelwold in the *Regularis Concordia* written in 963–75 for the use of Benedictine monasteries in England, it is a dialogue between one who sits with a palm-branch in his hand at the place of the sepulchre and three others who enter with censers in their hands, as if seeking something. And St. Ethelwold adds the significant gloss, 'Aguntur enim haec ad imitationem Angeli sedentis in monumento, atque mulierum cum aromatibus venientium ut ungerent corpus Ihesu'. Imitation; that is impersonation. In course of time the *Quem quaeritis*, now definitely an Easter play, received accretions. An episode was added for St. Peter and St. John; another for the risen Christ himself; a third, without scriptural warrant, in which the Marys buy their spices from an ointment-seller or merchant; a fourth for the Roman soldiers who fall senseless at the door of the sepulchre. Very elaborate versions of the thirteenth and fourteenth centuries are found in the so-called *Ludi Paschales* from Origny-Sainte-Benoîte, Klosterneuburg, and Tours, and in the *Carmina Burana*, probably written by wandering scholars and preserved at the monastery of Benediktbeuern. Meanwhile the *Quem quaeritis* had become the model for analogous dramas on other feast-days. On Easter Monday came a *Peregrinus*, dealing with the journey to Emmaus; at Christmas an *Officium Pastorum*, which centred round the crib and introduced the Midwives from the apocryphal *Pseudo-Evangelium Matthaei*; on Epiphany an *Officium Trium Regum*, also called *Officium Stellae*, from its use of the mechanical star. It was elaborated by a visit to Herod, soon described as 'wrathful', and this in turn led to a Slaughter of the Innocents, with Rachel weeping over her children, and a Flight into Egypt. The *Stella* also tended to absorb the

Pastores. This was perhaps due to the development at Christmas of another drama of rather different type. Its origin was not in a trope but in a sermon, the pseudo-Augustinian *Contra Judaeos, Paganos et Arianos Sermo de Symbolo*, in which the writer cites, as *testes Christi*, not only the prophets Isaiah, Jeremiah, Daniel, Moses, David, and Habbakuk, but also Simeon, Zacharias, Elizabeth, and John Baptist, and even the pagan Nebuchadnezzar, Virgil, and the Erythraean Sibyl. In the dramatic *Ordo Prophetarum* of Christmas these or some of them came forward and spoke for themselves, and a later development added Balaam and his Ass, thus introducing an element of comedy. The *Prophetae*, the *Pastores*, the *Tres Reges*, the *Innocentes*, are amalgamated in a play among the *Carmina Burana*, which also includes an Annunciation, not traceable on its own appropriate day until a comparatively late date, and a Visit to Elizabeth. From the *Prophetae* may have budded off an Isaac and Rebecca, a Joseph, and even more probably a lost Moses and a Daniel, of which there are two versions, one due to the twelfth-century wandering scholar, Hilarius. But there is no similar development of New Testament themes. A Passion play does not make its appearance until the thirteenth century, again in the *Carmina Burana*. But it may well have had its origin in certain lyrical *planctus*, or 'laments', which begin during the twelfth century and may originally have been attached to the Adoration of the Cross. They are mostly monologues of the Virgin at the Cross, but some are by Christ speaking from the Cross, and some are in dialogue by Mary and Christ, or Mary and St. John, or are spoken by the women at the sepulchre. It is not clear whether the *planctus* began within the liturgy itself or were borrowed from independent lyrical poems. From the feast-days of the main ecclesiastical cycle the instinct for drama spread to those of saints. We have two examples on the Raising of Lazarus, whose cult was on 17 December. One is again by Hilarius, another in a book of six plays from the monastery of St. Benoît-sur-Loire at Fleury. This also has one on the Conversion of St. Paul for 25 January. But the special patron of medieval clerks was St. Nicholas, and to his day on 6 December belong seven plays, of which one is by Hilarius and four come from the Fleury MS. Finally, there appear two elaborate plays, which may be called eschatological, a *Sponsus* on the theme of the Wise and Foolish Virgins, and an *Antichrist*. The liturgical relation of these, if any, is un-known. The Presentation of the Virgin on 21 November, the Purification on 2 February, and the Assumption on 15 August do not appear to have received dramatic treatment before the fourteenth century or later. The final development of the Marian cult was rather an affair of the later Middle Ages.

England probably had its full share in the symbolical cere-monies on festival days and the liturgical plays which grew out of them, although the destruction of service books at the Reformation must have somewhat blurred the recórd. Crowns for *representationes* are in a Salisbury inventory of 1222, stars for the *Pastores* and *Reges* in York *Statutes* of the thirteenth century. The Palm Sunday procession, the rending of the veil in Holy Week, the washing of feet on Maundy Thursday, are all well established. And on Palm Sunday at Salisbury and Wells appeared the prophet Caiaphas, who is not in any preserved version of the *Prophetae* and may have come direct from the Gospel of St. John. The Pentecost dove was still used at St. Paul's in the sixteenth century. The Visitation of the Cross prescribed by St. Ethelwold appears in the Winchester troper of slightly later date, and in elaborated versions from the Church of St. John the Evangelist at Dublin and the nunnery of Barking; but it cannot be shown to.have accompanied the Deposition and Elevation at Salisbury, York, Exeter, Hereford, or the Abbey of Durham. The statutes of Lichfield Cathedral, originally written at the end of the twelfth century and revised in the thirteenth or fourteenth, provide for representations of the *Pastores* at Christmas, the *Resurrectio* on Easter Day, and the *Peregrini*, or Pilgrims, on the following Monday. The last two they also describe as 'representations of miracles'. From William Fitzstephen we learn that in the twelfth century London had 'representationes miraculorum quae sancti confessores operati sunt, seu representationes passionum quibus claruit constantia martyrum'. A thirteenth-century sermon, from an unspecified locality, promises a play on St. Nicholas to come after it. How far the practice of cathedrals and monasteries was followed in the ordinary town and village churches we cannot say. The numerous structural sepulchres, often elaborately carved, which survive up and down the country suggest that the Easter rites at least were widespread. It may be that the play of St. Mary Magdalen and that of Christ and the women recorded in the accounts of Magdalen College, Oxford, in 1507 and 1519 respectively, were still liturgical; and possibly also the *Nativity* and *Resurrection*, for which rewards were provided, should they be performed, by the Earl of Northumberland at Leconfield in Yorkshire, about 1522. It is matter for regret that we have no texts of the ceremonies practised at the Cathedral of Lincoln, since the Chapter accounts from the fourteenth century onward suggest that these were both varied and often elaborate. Some of them seem to have recurred annually, others only on special occasions. The payments are generally to a porter or other official who provided the material requisites and also looked after the cathedral clock. During the fourteenth century the records show four performances of a *Tres Reges* at the Epiphany, two of a *Resurrectio*, doubtless on Easter Sunday, and five of a *Peregrinus* on Easter Monday. More puzzling is a *Salutatio* on Christmas Day in 1390, for which were provided a star and a dove. It has been suggested that this was a *Pastores*, but I am inclined to think that it was an *Annunciatio*, analogous to a contemporary one at Padua, which also had a dove, and in which a prophecy was spoken. The star does not properly belong either to a *Pastores* or an *Annunciatio*, but if such a fitting had been acquired for the Epiphany play, it might easily be used on other occasions. The term *Salutatio* elsewhere signifies the *Ave Maria* of Gabriel at the *Annunciatio*. The star, with the cords by which it was moved, sometimes occurs later in the accounts, and a Christmas *visio* of 1458 may have been a representation of the play. And it seems to have left its trace in the purchase, recorded in almost every year during a century and a half, of gloves at Christmas, which are first described as for Mary, Elizabeth, and the Angel, then for Mary, the Angel, and two Prophets, and then for Mary and the Angel alone. A visit to Elizabeth might naturally come into an *Annunciatio*, and the Prophets recall the prophecy at Padua. Perhaps one of them was Moses, since a *visio* called *Rubum quem viderat* is recorded at the Christmas of 1420. The dove was also used at Pentecost, and a banner was probably displayed during the singing of the hymn *Vexilla Regis prodeunt* on Passion Sunday. During the fifteenth century emerges an *Ascensio*, probably not of Christ but of the Virgin, and identical with the *Assumptio* or *Coronatio*, which now became a prominent ceremony at Lincoln. It took place, not at the feast of the Assumption on 15 August, but at that of St. Anne, the mother of the Virgin, on 26 July. In 1460 and 1469 there was a *visus Assumptionis*, and in 1483 an arrangement was arrived at with the city of Lincoln by which the *ludus sive seremonium de Assumpcione sive coronacione* should be incorporated with a civic procession on St. Anne's day. A new *mechanica coronacio* was made for this purpose. An analogous arrangement is found at Aberdeen, where a similar procession was attached to an offering of the Virgin at the Purification.

By the thirteenth century the liturgical drama, except for the few Marian additions, had reached the full term of its natural development and was tending to overshadow the sacrificial and devotional ritual of the feast-days, which it was originally intended to adorn. Probably this was largely due to the literary activities of the wandering scholars, with their tendencies to realism and even hilarity. The Benedictine monasteries harboured the plays, but the reformed orders remained aloof. Some of the austerer clergy became critical. Gerhoh of Reichersberg condemns the monks of Augsburg, who never made an appearance in their house, unless a performance of Herod or some other theatrical spectacle gave occasion for a banquet. He is echoed in this country by Aelred of Rievaulx, who regrets the introduction into the liturgy of histrionic gestures, more suitable for a theatre than an oratory. In the more highly developed plays two notes of transition may be traced. The strict ecclesiastical Latin was gradually invaded by elements of the vernacular. Many of the Easter plays from Germany end with a hymn *Christ ist erstanden*, to be sung by the congregation. Hilarius and others were fond of glossing their Latin speeches with paraphrases or even substantial additions in German or French. The *Ludus Paschalis* of Origny-Sainte-Benoîte and the *Sponsus* have scenes or passages entirely in French. England contributes the bilingual fragments found in a manuscript at Shrewsbury, although this itself is not of earlier date than the sixteenth century, and probably therefore preserves a survival. It contains the 'parts', with their cues, of a single actor in a *Pastores*, a *Visitatio Sepulchri*, and a *Peregrinus*, which may have been given either at Shrewsbury itself or in the mother church of Lichfield. The dialogue seems to have been mostly in English, with Latin passages interspersed. As transitional we may also regard the growing multiplication of episodes, within the limits of a single performance. It was no longer possible for these to be confined to the choir, where the liturgical ceremonies themselves took place. Their action had to extend down the nave, with platforms, or *sedes*, for the successive scenes ranged against its pillars, leaving the sepulchre perhaps in its original position, while a hell might stand, appropriately enough, either at the west door or by the steps leading down to a crypt. It is not even certain that some of the more elaborate Latin plays may not have been performed in front of rather than inside a church, but of this there is no definite proof. At some time or other, however, this change came about, at any rate in certain localities. During the winter of 1204 we hear of a *ludus prophetarum ornatissimus* which took place *in media Riga*. It had an *interpres*, which I take to mean an expounder, rather than a translator. At Padua, in 1244, a Passion was represented, apparently in the open. From England comes a record of a Resurrection produced about 1220 in the churchyard of Beverley Minster, which was sanctified by the miraculous restoration of a boy who had climbed up to a window inside the church to get a better view and had fallen on the pavement. In the open, although drama continued to be religious in subject-matter, its divorce from the liturgy was completed. Vernacular, once an intrusion, became its regular medium. Its organization and finance passed into the hands of civic authorities or other bodies of a lay, rather than an ecclesiastical character, although doubtless it continued to look to clerical writers for the provision of texts. In sixteenth-century Scotland religious plays were still called 'clerk plays'. In a churchyard, a market-place, a meadow, the amalgamation of themes could receive a further extension. Seasonal conditions made this almost inevitable. An out-of-doors play in winter might be possible under the sunny skies of Italy, but not on the icy shores of the Baltic at Riga or in this island. The Christmas and Easter themes were combined in a single performance, and by prefixing a Creation and Fall, already traceable at Regensburg in 1195, and expanding the hint of such a play as the *Sponsus* into a *Judicium*, a type of cyclical drama was arrived at, which was coextensive with the span of humanity itself, as medievalism conceived it. The process was complete at Cividale in Friuli by 1304, and in England perhaps not much later.

But here, as the liturgical books fail us, we enter upon an obscure period of transition, which lasts well into the fourteenth century. A piece on the Harrowing of Hell, once thought to be a play, is now regarded as no more than a narrative *estrif* or *débat*, intended for recitation. Three dramatic texts survive, all of which are fragmentary. They are not in English but in Anglo-Norman French. It is probable, however, that they belong to this country. The earliest, which may date from the middle of the twelfth century, is headed *Ordo representacionis Adae*. It is still not quite free from liturgical connexion, since it has a *Chorus* and begins with a *lectio*, probably the pseudo-Augustinian *Sermo de Symbolo*. But it was evidently staged outside a church, from and to which the stoled *Figura* of the Creator Himself moved. The *plateae*, about which ran demons, held a raised 'paradise', with a 'forbidden tree', an artfully constructed serpent, and a hell. The fragment includes a Creation and Fall, a Cain and Abel, and an incomplete *Prophetae*, with a Balaam and his ass. The whole performance must have been much like that at Riga in 1204, but there is no mention of an *Expositor*. The second Anglo-Norman text is of a *Resurrection*. Of this there are two versions. One has long been known. The other, recently found in a manuscript long preserved at Canterbury and datable about 1275, has not yet been published in full. Rather curiously, the octosyllabic couplets, in which the dialogue is written, are used also for the stage directions, and for a prologue. The action, so far as the fragments go, includes the request of Joseph and Nicodemus for the body of the Saviour, the setting of a watch at the sepulchre, the arrest of Joseph and his examination by Caiaphas. But the prologue also contemplates a Harrowing of Hell, an apparition to the Maries, and a *Peregrinus*. The setting required a crucifix, a monument, a jail, a heaven, a hell, and platforms to indicate Galilee and Emmaus. The third Anglo-Norman text is a very short one, merely a scrap in French with an English translation, containing a speech by a King, who might be Octavian or Herod, to a Messenger. On the back of the manuscript is a memorandum about a manor once held by the Abbey of Bury St. Edmunds, which suggests the possibility that we may here have a fragment from a play given at that house. There is no reason to infer from these survivals that the habit of playing in the vernacular was an English borrowing from France. The earliest French *Passion* plays are in fact of later date than the Anglo-Norman *Resurrection*. They are probably based upon a narrative poem known as the 'Passion des Jongleurs', and of this also there is an Anglo-Norman version, which may have been used by the author of the *Resurrection*, and also by later English playwrights.

Dr. Owst has rightly called attention to the frequency with which topics dealt with in the plays are anticipated in the *exempla* and the homely illustrations of popular preachers. Nevertheless the vernacular play did not establish itself in this country without incurring some theological disapproval. About 1244 Robert Grosseteste, Bishop of Lincoln, wrote to the archdeacons of his diocese bidding them to exterminate the *ludos quos vocant miracula* performed by clerks, and classing them with the *scotales* or drinking bouts and the seasonal revels in spring and autumn of the folk. He was in a position to dictate without giving his reasons. But a clear discrimination between ritual plays in church and performances out of doors is made about 1300 in the *Manuel des Pechiez* of William of Wadington, who writes:

> Vn autre folie apert
> Vnt les fols clercs cuntroué,
> Qe 'miracles' sunt apelé;
> Lur faces vnt la deguisé
> Par visers, li forsené,
> Qe est defendu en decrée;
> Tant est plus grand lur peché.
> Fere poent representement,
> Mes qe ceo seit chastement
> En office de seint eglise
> Quant hom fet le Deu seruise,
> Cum Ihesu Crist le fiz dée
> En sepulcre esteit posé,
> E la resurrectiun,
> Pur plus auer deuociun.
> Mes, fere foles assemblez
> En les rues des citez,
> Ou en cymiters apres mangers,
> Quant venent les fols volunters,
> Tut dient qe il le funt pur bien,
> Crere ne les deuez pur rien
> Qe fet seit pur le honur de Dée,
> Einz del deable, pur verité.
> Seint Ysidre me ad testimoné
> Qe fu si bon clerc lettré;

Il dist qe cil qe funt sepectacles
Cume lem fet en miracles,
Ou ius qe nus nomames einz,
Burdiz ou turneinens,
Lur baptesme vnt refusez,
E Deu de ciel reneiez.

A translation may be helpful here.

Foolish clerks have devised another open folly, which they call 'miracles'. Their faces are disguised by masks, the madness which is forbidden by decree. So much the greater is their sin. They may make a representation, if it is done modestly in the office of Holy Church, when man renders service to God, of how Jesus Christ, the Son of God, was laid in the sepulchre, and of the Resurrection, for the sake of greater devoutness. But if they make foolish assemblies in city streets or in graveyards after meals, when fools are glad to come, even though they say that it is for a good purpose, do not on any account believe them that it is done for the honour of God, but rather in truth for that of the devil. St. Isidore, that well-lettered clerk, is my witness. He says that those who make shows, as is done in other sports called miracles, or in tilting or tournaments, have refused their baptism, and denied the God of heaven.

Wadington is paraphrased in 1303 by Robert de Brunne in his *Handlyng Synne*:

Hyt is forbode hym, yn the decré,
Myracles for to make or se;
For, myracles ȝyf thou bygynne,
Hyt ys a gaderyng, a syght of synne,
He may yn the cherche thurgh thys resun
Pley the resurreccyun,—
That ys to seye, how God ros,
God and man yn myȝt and los,[1]—
To make men be yn beleuë gode
That he ros with flesshe and blode;
And he may pleye withoutyn plyght[2]
Howe God was bore yn ȝolë[3] nyght,
To make men to beleue stedfastly
That he lyght yn the Vyrgyne Mary.
Ȝif thou do hyt yn weyys or grenys,
A syght of synne truly hyt semys.
Seynt Ysodre, y take to wyttnes,
For he hyt seyth that sothe[4] hyt es;

---

[1] los (loss), 'defeat'.   [2] plyght, 'offence'.   [3] ȝole (Yule), 'Christmas'.
[4] sothe (sooth), 'truth'.

Thus hyt seyth, yn hys boke,
They forsake that they toke—
God and here crystendom—
That make swyche pleyys to any man
As myracles and bourdys,[1]
Or tournamentys of grete prys.

The distinction here drawn is clear enough, although the authorities cited are grievously misinterpreted. The 'decré' is a letter of Innocent III in 1207 against the 'theatrical' use of masks in churches and the seasonal Feasts of Fools at Christmas, which was included by Gregory IX in his *Decretals* of 1227–41; and Isidore of Seville, about 636, was condemning nothing more relevant than the spectacles of the Roman stage. This early criticism of the popular drama seems to find an echo about 1360 in the *Summa Predicantium* of the Dominican John de Bromyard, who more than once refers to the plays called *miracula*, and under the heading *Audire Verbum Dei* complains of those who seem to have plenty of time to witness them, but say that their business prevents them from listening to sermons. This note of hostility does not recur in the writings of later orthodox moralists, but it is taken up about the end of the fourteenth century in an elaborate Wycliffite *Tretise of Miraclis Pleyinge*. This is unsparing in its condemnation, rejecting the argument that the performances are done in the worship of God and move the audience to compassion and devotion, as well as the weaker plea that men must have some recreation and better 'by pleyinge of myraclis than bi pleyinge of other japis'.[2] Those who take the most precious works of God 'in pley and bourde',[3] says the writer, are scorning him, 'as dyden the Jewis that bobbiden[4] Crist'. To the author, on the other hand, of *Dives et Pauper* (*c.* 1410), plays are lawful if they are performed without 'ribaudry' or 'lesynge',[5] and at times which do not interfere with 'goddes servyce'.

Scholars are not wholly agreed as to the name by which a vernacular religious play was known in the thirteenth and early fourteenth centuries. No doubt it was often simply called a 'play'.

Yf ye wollet stillen ben,
In this pleye ye mowen isen,

---

[1] bourdys, 'farces'.   [2] japis, 'jests'.   [3] bourde, 'farce'.   [4] bobbiden (bobbed), 'struck with the fist'.   [5] lesynge, 'lying'.

says the homily which preceded a performance of about 1250.
But 'play', like the Latin *ludus*, is wide enough to cover any
kind of amusement. It might be called an 'interlude'. 'How
thanne may a priest play in entirlodies', asks the author of the
Tretise. An interlude is a play 'between' speakers, a play in
dialogue. But it is not necessarily religious. An *Interludium de
Clerico et Puella* of the end of the thirteenth century is a farce.
The liturgical texts give us little help here. The rubrics which
introduce their dramatic passages follow no uniform practice.
They may use a formula which is itself purely liturgical. The
passage is an *ordo, officium, caeremonia, ritus*, or *solemnitas*. They
may merely give it a subject-title, such as *Visitatio Sepulchri*.
On the other hand, they may show a consciousness that it
involves impersonation. It is a *similitudo*, an *exemplum*. But
the commonest of such terms and the one which is most clearly
of literary rather than of liturgical significance is certainly
presentatio or *representatio*. This survived in the Italian vernacular
as *rappresentazione*. The Lincoln *visus* is of late date. I feel little
doubt myself that the normal English term was 'miracle'. It is
used both by Grosseteste and by Bromyard, in a latinized form,
but with wording which makes it clear that they are translating
from the vernacular. Nor do these passages stand alone. They
are echoed in the thirteenth century by Matthew Paris, who
tells of a *ludus* of St. Catherine, *quem miracula vulgariter appellamus*,
given by Geoffrey of Gorham in the twelfth, and in a Latin
story of two Franciscans, who were travelling near Corinth in
1266 and there saw a crowd in a meadow, now silent, now
acclaiming, now *cachinnantes*, whom they judged to be witnessing
spectacula quae nos miracula appellare consuevimus. This interpretation of the term 'miracle' as covering all out-of-door vernacular
religious plays is consistent with the language of William of
Wadington, who draws a sharp distinction between a legitimate
representement in church and an irregular *spectacle* in a street or
graveyard, such as the 'miracles' provided. 'Spectacle' itself,
which the moralists sometimes parodied as 'steracle', is of
course again of the widest significance. It has, however, been
argued that the term 'miracle' does not cover all religious plays,
but only those of which the subject-matter was drawn from the
legends of saints. This view seems to rest in part on the fact
that Geoffrey of Gorham's play was on St. Catherine, and in
part on the rubrics to one of the Fleury liturgical plays, which

is headed with *Aliud miraculum de Sancto Nicholao* and ended with
Finitur miraculum, although I do not see clearly why these need
be more than indications of the subject-matter. But in the main
I think it is an illegitimate transference to English usage of a
distinction between hagiological *miracles* and biblical *mystères*,
which was made in French terminology from the end of the
fourteenth century onwards. No English play was, however,
called a 'mystery' before 1744. Supporters of the theory sometimes cite the statement by William Fitzstephen about 1170–82,
on plays in London, which has already been quoted. Here,
however, the technical term is not *miraculum* but still the liturgical *repraesentatio*. Certainly the writer of the *Tretise of Miraclis
Pleyinge* was not thinking primarily of saints' plays when he
condemned the playing of 'the myraclis and werkis that Crist
so ernystfully wrou3te to our helye',[1] and more specifically
'the pley of Cristis passioun', and that 'of Anti-Crist and of
the day of dome'. Nor can we suppose that, when Chaucer's
Wife of Bath amused herself during her husband's absence by
going to 'pleyes of miracles', it was saints' plays only that
attracted her.

Miracle plays are traceable during the fourteenth and
fifteenth centuries in some forty English localities, predominantly perhaps in the northern and eastern parts of the
country. Others can be added from sixteenth-century documents. They were known also in Scotland, and at Dublin in
Ireland, which was much under the influence of English
customs. The fact that most of the surviving texts represent
performances of a particular type has perhaps rather obscured
the variety of organization which the records disclose. It is
probable that, when the plays were detached from the liturgy,
the financial responsibility was often taken over from the
established clergy by some of those religious and social gilds
which contributed so largely to the development of medieval
public life. In London this arrangement seems to have prevailed to the end. Here a gild of St. Nicholas, composed of
parish clerks, gave performances, possibly with the assistance
of boys from the cathedral school of St. Paul's, in the open at
Skinners Well, near Clerkenwell in the suburbs. The plays are
first noticed in 1300, when the Abbess of Clerkenwell com-
plained to the king of the damage done to her fields and crops

[1] *helye* (heal), 'health'.

by the crowds who attended the 'miracles', as well as the wrestling bouts which the citizens were accustomed to hold in the same locality. The plays, however, continued. Stow says that they were given annually. The chroniclers record elaborate performances in various years from 1384 to 1508, lasting from four to seven days, and covering the whole span from the Creation to the day of Judgement. In 1508, however, they seem to have been given near St. Paul's. They took place in summer, twice at least on a day dedicated to St. John Baptist. London, however, had also, as in the days of William Fitz-stephen, saints' plays. There was one on St. Catherine in 1393, and others at St. Margaret's Church, Southwark, on the feasts of St. Margaret and St. Lucy, in the fifteenth century. Outside London, SS. George, Thomas of Canterbury, Laurence, Diony-sius, Susanna, Clara, and no doubt many others, were similarly commemorated. Some of these plays seem still to have been given in churches, and others are recorded in churchwardens' accounts well into the sixteenth century. Presumably, however, they were now detached from the liturgy, and the sacred edifice merely served to shelter them. At Hedon, in Yorkshire, in 1391, we find the town chamberlain making a payment to Master William Reef and his companions for playing on Epiphany morning in the Chapel of St. Augustine. Probably these companions were a gild. There are occasional notices also of plays in a gild hall, or by schoolboys at the manor-house of some person of local importance. But in the main miracle plays were an out-of-doors affair. Many of the small religious and social gilds disappeared during the Black Death, and by the fifteenth century the control of the plays, in the larger towns outside London, seems generally to have passed to the local governing body, itself often in some aspects by origin a gild of the merchant gild type. At Norwich, however, the gild of St. Luke remained responsible for the plays at least to 1527. Many of the performances seem to have been occasional only, given in the market-place or in some convenient 'croft' or 'stead'. At Aberdeen they were on Windmill Hill; at Lincoln in the cathedral close. More interesting, from the literary point of view, is the emergence of a type of play which was not stationary but ambulatory, taking the form of a procession up and down the streets, with pauses here and there for the repetition of scene after scene at traditional spots. So used, the plays seem to have become, like the simpler processions of the Rogation days, popularly called Gang Week, a sanctification of the old perambulations of the pagan cults which took place at critical seasons of the agricultural year, in winter when the ploughing began, and again in the heats of summer when the crops were growing. At Hull, on Plough Day, a play of Noah was given on a stage in the form of a ship, which was also carried about the town in a ceremony which clearly represents a maritime version of the agricultural rite. Normally, however, the processional miracle plays took place in summer. At Chester and Norwich, where an early practice seems to have been preserved, they were at Whitsuntide. It is not clear whether Whitsun plays at Leicester and New Romney were processional or stationary. But elsewhere a new date was provided by the papal confirmation in 1311 of the feast of Corpus Christi, tentatively contemplated in 1264. This fell in May or June and was essentially a processional observance, in which the Host itself was carried around and displayed with ceremony at appointed stations. And to it in various ways the miracle plays often became attached. Many gilds of Corpus Christi were founded. One at Bury St. Edmunds, which claimed in 1389 to have existed from time beyond memory, was bound by its constitution to honour the occasion with an interlude. Corpus Christi plays now become traceable all over the country, but perhaps predominantly in the north and east, during a period of more than two centuries. They are found also in Scotland. At Aberdeen they were Haliblude plays. The Reformation hit them hard, but a last survival is recorded at Kendal, about 1612. Some of the performances were stationary and perhaps occasional only. But in many large towns, including most of those from which the surviving texts come, the processional type prevailed. It was so at York, Wakefield, Beverley, Doncaster, Newcastle, Ipswich, Coventry, and Worcester. The organization seems to have been much the same in all these places so far as the records enable us to judge. It was adopted also at Chester during the period which our information covers, although here the Whitsun date was retained. A general control was exercised by the council of the town. Subject to its approval, which would naturally be withheld in times of pestilence or public disturbance, the plays were given annually. The actual performance of them was still in the hands of gilds. These, however, were not of the old type, but were trade or craft gilds which, while still often retaining a religious and social side, were primarily organizations for the promotion and regulation of local industries. A cyclical theme, which usually extended from the Creation to the Last Judgement, was divided into a series of scenes, for the production of each of which a particular gild took responsibility. The performance was given on large wagons, which followed each other in regular sequence to station after station about the town, at which scene after scene was repeated. These wagons were known as pageants. The origin of this term, which was also often used as an alternative to 'play' for the scenes themselves, is not quite clear. It may be derived from a Latin *pagina*, either in the sense of the page of a book or in that of a framework. There was some superstructure representing a *domus* or *sedes* for the actors, with a hell and the like, but it must have been rather sketchy, to allow of a view from all sides, and we know little about it, beyond a late Chester description which says that the performers had 'a highe place made like a howse with ij rowmes, beinge open on ye tope: in the lower rowme they apparelled & dressed them selues; and in the higher rowme they played; and they stoode vpon 6 wheeles'. The cost of maintaining and housing the pageants and of paying the actors fell upon the crafts, each member of which was bound to make an annual contribution for the purpose. This involved a good deal of work for the governors, as industries rose or fell and claimed to have or be relieved from a share in the performances. Such changes are often recorded in town documents or have left their marks on the structure of the surviving texts. These are mainly of late date and preserved in registers kept by the corporations, themselves compiled from 'originals' in the hands of the crafts. The plays flourished. *Vexillatores*, or banner-bearers, rode about in advance, reading banns which announced the subject-matter of the scenes. Spectators thronged in from neighbouring villages. Corpus Christi day became a great public holiday, which brought much profit to a town, although perhaps more to its hostelers and victuallers than to the productive trades on which the main financial burden fell. As a result, the religious procession tended to become overshadowed. This was sometimes met by shifting the plays, or even the procession itself, to a neighbouring day. But elsewhere, for example at Dublin, we find no more

than a symbolical riding with the Host of personages taken
from biblical or legendary narratives, perhaps formerly the
themes of plays given in full. At Lincoln, on St. Anne's day,
at Beverley on that of the Purification, and at Aberdeen on
Candlemas, there were similar ridings which seem to have
ended with plays, possibly still liturgical, in the churches. The
enduring human instinct for *mimesis* is capable of many mani-
festations. It must be added that in course of time the term
'Corpus Christi play' seems to have been applied to any
representation of the Passion and Resurrection, at whatever
date it was given. It was so at Chester, where the civic plays
seem to have been always at Whitsun, although in the middle
of the sixteenth century the colleges and priests had an independent one of their own on Corpus Christi day. At Lincoln
in the fifteenth century there were occasional plays, apparently
distinct from the procession on St. Anne's day, among which a
ludus de Corpore Christi took its turn with others. When the
governors of a town, in the fourteenth century, wanted to
establish a dramatic cycle, they would naturally turn for a text
to some local ecclesiastic, able to compose in English metre and
free from theological prejudices against miracle plays. He
might be a gild priest, or more probably, in a large place, a
brother from some monastery, such as St. Werburgh's at
Chester. The tradition of the liturgical drama would be behind
him. And for its expansion he would have a considerable
library at his disposal. He could gloss the narrative of the Bible
itself with much legendary material that had grown up around
it throughout the ages. Perhaps he would not have read the
earlier works from which this took its start, such as the *Vita
Adae et Evae*, the *Evangelium Nicodemi*, which comprised the *Acta
Pilati* and the *Descensus Christi ad Inferos*, the *Evangelium Matthaei*,
with its *De Nativitate Mariae*, the *Transitus Mariae*. But he would
certainly be familiar with some of the great medieval compilations, such as the *Allegoriae in Vetus Testamentum* of Hugo de
St. Victor, the *Historia Scholastica* of Peter Comestor, the *Legenda
Aurea* of Jacobus de Voragine, the *Estoire de la Bible* of Hermann
de Valenciennes. He might know the *Meditationes* on the life
of Christ ascribed to St. Bernard of Clairvaux and St. Bonaventura of Padua, as well as the pseudo-Augustinian *Sermo de
Symbolo* and the *De Antichristo* of Adso of Toul. There were
vernacular poems, too, on which some of these sources had

already had their influence, the Middle English *Harrowing of Hell*, the *Genesis and Exodus*, the *Cursor Mundi*, the *Northern Passion*, based on the *Passion des Jongleurs*, a similar *Southern Passion*, the *Life of Saint Anne*, the so-called *Stanzaic Life of Christ*, an English *Gospel of St. Nicodemus*, versions of the Holy Rood legend, and many lyrical poems, some of them of the *planctus* type. There was abundance of material, instructive and entertaining, to draw upon. Some of it, however, might not be available until well into the fourteenth century. The probability that some cycles of late origin may have borrowed or adapted plays from others already in existence must also be kept in mind. No doubt the earliest writers aimed primarily at edification. But the folk, on a holiday, had also to be entertained. They could be moved by the tragedy of the divine sacrifice, and perhaps exalted by the scheme of creation and redemption. But they found their relief in episodes which involved an element of humour, in the unwillingness of Noah's wife to enter the Ark, in the homely banter of the shepherds before the angel came, in the ranting of the potentates, Octavian, Herod, Pilate, in the 'bobbing' of the Redeemer by the torturers, in the dice-play of the soldiers over his garments, in the downfall of the tax-gatherers and other unpopular elements of society at the Last Day. Sometimes a play became so farcical in action, if not in language, that it had to be pruned or abandoned. At York the midwives were cut out of the Nativity, and the antics of Fergus at the bier of the Virgin provoked such unholy mirth that the responsible craft became unwilling to repeat them. It must be added that the texts of such plays as survive have come down to us in a very corrupt state. The earliest are of the middle of the fifteenth century; others are as late as the end of the sixteenth. Some have been completely rewritten. The older ones have often been patched in metres incongruous with those in which they were originally written. Much corruption has also been introduced by the characteristic unreliability in transcription of medieval scribes, often increased by blundering attempts to reproduce old linguistic forms which had become unintelligible. Many single lines and even longer passages have evidently been omitted, to the detriment both of sense and once more of metre. Whoso would read the plays to-day must often go darkling.

Of the many craft-plays, whose existence we can infer from records, edacious time has left us comparatively few examples.

We have cycles, practically complete, from Chester, York, and Wakefield, two very long plays from Coventry, one from Newcastle, one from Norwich, and one other, the origin of which is unknown. The loss of the Beverley plays, which existed in 1377 and were already an ancient custom in 1390, and about which we have many details, is particularly regrettable, since they might link with the churchyard representation of a Resurrection about 1220.

Probably the earliest surviving cycle, at least by origin, is that from Chester. On this we have abundant material, but unfortunately it is nearly all of very late date, and has in part come to us through the hands of writers who blended an antiquarian interest in the history of their town with a Puritan dislike of the ecclesiastical tradition to which the plays belong. Performances, probably becoming no longer annual, had continued during the sixteenth century. The latest upon record were in 1572 and 1575, and both of these brought the mayors under whom they were given into trouble with the diocesan authorities and the Privy Council. An alleged revival in 1600 is likely to rest upon a misunderstanding. We have five texts of the cycle, dated from 1591 to 1607. According to Dr. W. W. Greg, who has carefully studied their interrelation, they are all ultimately derived, through one or more intermediate transcripts, from an official register, perhaps of the later half of the fifteenth century. One, however, in the Harleian collection, although the latest in date, seems to rest upon an earlier version of the register than the others. They include all the plays of which we have any record, except an Assumption, which existed in 1500 and 1540, but may never have been registered. We have also a separate text of the Antichrist, apparently from a prompter's copy, written about 1500, and late ones of the Trial and Flagellation from an original, and of a fragment of the Resurrection. We have two sets of banns, which record the crafts and the subjects of their plays. One is from an official document of 1540, but the dislocation of its metre by the incorporation of changes shows that it must have been by origin of much earlier date. The other probably represents a performance which may be as late as 1572. But in its fullest form it has been much glossed with side-notes of Protestant comment, which may be due to one Robert Rogers, an archdeacon and prebend of Chester, who died in 1595. A final

passage has been thought to suggest that an indoor representation was at one time contemplated. But this must remain very
doubtful. Finally there are several copies of a late list of plays,
showing them as twenty-five in all, but not including the
Assumption, and indicating their distribution over the three
days of performance. It is certain that during the course of
over two centuries, through which the plays lasted, they had
grown in number. There were only twenty or twenty-one
about 1475, and twenty-four about 1500. Professor F. M. Salter
has attempted to trace an earlier development by exploring
craft records for indications of changes of responsibility, which
can be correlated with the metrical irregularities of the early
banns. As a result he infers that those banns were originally
composed about 1467, and that the cycle then consisted of only
eighteen plays. A discussion of his argument in detail would be
inappropriate here, but some such development is likely enough.

One change, of which there is a fairly full record, may be
cited, both for the light which it throws upon the handling of
contentious citizens and because it involves something of a
puzzle. In 1422 a dispute arose between the Ironmongers and
the Carpenters or Wrights, both of whom claimed to have the
assistance of a group of small crafts, headed by the Coopers,
in the production of their plays. It came before the portmoot,
who referred it for arbitration to a jury. As a result it was
decided that neither contention should be accepted, but that
the Coopers and their fellows should have an independent play
of the Flagellation of Christ, while the Ironmongers should take
the Crucifixion, which immediately followed it, and the Wrights
should similarly have their own play, which, at any rate later,
was the Nativity. It seems to have been long before this
arrangement was recorded in the register, and then the method
adopted was a very rough one. In the Harleian MS. the Trial
and Flagellation and the Crucifixion still form one play of the
Passion, which is ascribed to the Ironmongers. In the other
cyclical manuscripts they are also written as one play, but
inserted stage-directions and additional lines show that in fact
they were meant to be played as two. They are two also in all
the late lists. The puzzle arises from the fact that the late banns
once more treat them as one. This has led to a theory, espoused
by Dr. Greg, that the change of 1422 was not a division but an
amalgamation. It seems to me inconsistent with the terms of

the arbitration, and also with the existence of a separate copy
of the Trial and Flagellation alone, preserved by the Coopers
Gild and dated in 1599. I take it that the writer of the late
banns was misled by the Harleian MS. or some other transcript
of the unrevised register.

The twenty-five plays of the Chester cycle, as we have them,
cover the whole range of the divine scheme for humanity.
Pre-Christian history is represented by a Fall of Lucifer, a
Creation and Fall, a Noah, a Cain and Abel, an Abraham and
Melchisedek, an Abraham and Isaac, a *Prophetae*, with Moses,
Balaam and Balak, and Balaam's Ass. A Nativity group includes
an Annunciation, a Visit to Elizabeth, Joseph's Trouble, further
prophecies of Sibylla to Octavian, a Nativity with midwives,
a *Pastores*, a *Magi* and Herod, with yet more prophecies, an
Oblation of the *Magi*, a Slaughter of the Innocents, a Death of
Herod, a Purification. To the missionary life of Christ belong
a Disputation with the Doctors, apparently a late addition and
borrowed from York, a Temptation, an Adulteress, a Healing
of the Blind, a Raising of Lazarus. The Entry into Jerusalem,
the Cleansing of the Temple, the Visit to Simon the Leper, and
the Conspiracy of the Jews with Judas are dealt with, rather
briefly, in a single play. The Last Supper and Betrayal at
Gethsemane occupy another. Then come the Passion proper, a
Harrowing of Hell, the Resurrection with the *Quem Quaeritis*,
a *Peregrini*, the Ascension, the Pentecost. The legends of the
end of the Virgin are untouched. But eschatology contributes
the Prophets of the Day of Doom, an Antichrist, and the Day
of Doom or *Judicium* itself. The sixteenth-century tradition, in
its earliest form, assigned the origin of the plays to the time of
John Arneway, who was mayor during 1268 to 1277, and
apparently the actual writing of them to Henry Francis, a
monk of St. Werburgh's Abbey in Chester, who was said to
have obtained an indulgence from the Pope for all who beheld
them in peace and devotion. Later notices substitute Randulph
Higden, also of St. Werburgh's, as the author. These statements
do not hold together. Higden, best known by his *Polychronicon*,
took the vows in 1299 and died in 1364. Francis was senior
monk of the abbey in 1377 and 1388. I have suggested elsewhere that Arneway may have been confused with Richard
Erneis or Herneys, who was mayor from 1327 to 1329. This is
not an impossible date for the initiation of a processional cycle.

The ascription to Higden, however, is not very plausible. His was, no doubt, a famous name in the annals of the abbey. But all his known or suggested writings are in Latin, and in the *Polychronicon* he tells us that in his day English *in paucis adhuc agrestibus vix remansit*. He may have exaggerated, but probably his own vernacular was Norman French. The metre of the Chester plays differs from that of the other dramatic cycles preserved to us by its uniformity. It is a Romance metre of the type known as *rime couée* or tail-rhyme, written normally in eight-lined stanzas, in which two *pedes* of three four-stressed lines rhyming together are each followed by a three-stressed *cauda*. The *caudae* rhyme with each other. The *pedes* may have a common rhyme, but more often do not. The rhythm is normally iambic. There is little alliteration, except in occasional passages, which may have undergone revision. The technical formula for the metre is $aaa_4b_3ccc_4b_3$ or $aaa_4b_3aaa_4b_3$. Obviously it is of a rather lyrical character, better adapted to romantic narrative than to drama, since it does not lend itself well to the quick exchange of speech and reply which dialogue often requires. But it is capable of some modification to suit changes of theme or tone or speaker, by reducing the number of lines in the *pedes*, or the number of stresses throughout. The inter-vention of a supernatural speaker, for example, is sometimes so marked. The early banns appear to have been originally written in the same stanza as the plays. It has been suggested that this metrical uniformity points to a wholesale rewriting of the plays at some stage in their history. I do not see any reason for this. What really wants explanation is the variety of form found elsewhere. Certainly the Chester stanza would have been available for a playwright as early as 1327; it had already been used for narrative romances in the north. It must be added that, in some of the plays, the original metre has certainly been patched with others by later hands. There are bits in couplets, in quatrains and octaves of cross-rhyme, variously stressed, even in rhyme-royal and its probably earlier four-stress equivalent. These signs of revision are particularly noticeable in the Fall of Lucifer and the *Pastores*. The episode between Christ and the Doctors, which was apparently borrowed from York and attached rather inappropriately to the Purification, is in cross-rhyme. A speech of Christ at the Resurrection, in a rather unusual form of cross-rhyme, with alternate four-stressed and three-stressed lines, may possibly be taken from some contemporary lyric poem.

I have an impression that behind the Chester cycle, as it has come down to us, lies a play of a more primitive type, the themes of which have been rather clumsily incorporated, with the result that discrepancies have been left in the action, which the late transcribers of the text have variously attempted to remove. The influence of the old play is clearest in those scenes in which an *Expositor*, also called *Preco*, *Doctor*, *Nuntius* or Messenger, comments to the 'Lordinges' of the audience on the significance of the topics represented. He calls himself Gobet on the Grene, and his demands for 'room' to be made, with the fact that both he and later the character *Antichristus* come in riding, suggest a stationary performance on a green or other open space, rather than one on moving pageants. The appearances and speeches of the *Expositor* indicate that the primitive play contained at least a Noah, an Abraham, a *Prophetae*, with Moses, Balaam and Balak, and Balaam's Ass, a Nativity, which was given on a second day and brought in the *Magi* as well as Octavian and Sibylla, and a Prophets of Doomsday. Probably it had also both a Doomsday itself and an Antichrist. The *Expositor* does not appear in the extant texts of these, but he foretells them. There is nothing to suggest any treatment in the primitive play of the missionary life of Christ or of the Passion and its sequels up to Pentecost. It is true that the *Expositor* is also in the Temptation and Adulteress episodes, but here I think he may have been borrowed by a later writer. The whole emphasis of the primitive play seems to have been on prophecies of the coming of Christ and of the Last Judgement. And here it fell into line with such earlier work as the Tegernsee *Ludus de Antichristo* and the Anglo-Norman *Adam* of the twelfth century, the Benediktbeuern and Riga *Prophetae* of the thirteenth, and the Rouen *Prophetae* of the fourteenth. Its *Expositor* looks very much like the Riga *interpres*. It might have been written as late as the fourteenth century, either in Latin or in French or in English, and if in Latin or French, it is conceivable, I suppose, that it might have been the work of Higden, although I attach little importance to the traditional ascription of the Chester plays to him. As to other possible sources available for a Chester playwright, little can be said. There was plenty of Latin material to be drawn upon. A phrase or two of French for a potentate is an insufficient basis on which to establish a connexion with any continental *mystère* known to us. Nor has any use of the fourteenth-century English *Cursor Mundi* been clearly demonstrated, although that encyclopaedic work was itself of northern origin. The so-called *Stanzaic Life of Christ*, apparently written at Chester in the first half of the same century, and in part at least based upon Higden's *Polychronicon*, may have contributed something.

It is difficult to arrive at any very clear estimate of the literary value of plays which were never intended to be read, and cannot be given life by the gestures and intonation of the actors. The Chester plays retain some attractiveness through their lyrical form, where that has not been blurred by the activities of revisers and scribes. Broadly speaking, they preserve more than the other vernacular cycles of the devotional impulse, which brought medieval drama as a whole into existence. Their preoccupation with prophecies, whether read into Old Testament narratives or derived from legendary sources, of the Redemption and the Day of Doom has already been made clear. They are also much concerned with the exposition of religious formularies, such as the Ten Commandments, and of Jewish and Christian ritual observances, such as the Sabbath, Circumcision, or the feast of First Fruits. The dramatic action is generally simple and straightforward, without much attempt to exploit the psychological possibilities of the themes dealt with. The human element, for example, in the relations between Joseph and Mary, which gives so much interest to the later cycles, is here little elaborated. There is some pathos, however, in the Abraham and Isaac play. The fundamental inhumanity of that story was, of course, not apparent to our medieval ancestors. The element of farce has as yet hardly obtruded itself. Lucifer takes his downfall with comparative submission. Cain is restrained in his attitude towards his brother. Balaam's Ass is not unduly exploited. The midwives in the Nativity remain decent. Noah's wife is already a little recalcitrant when she is asked to enter the Ark. She must have her gossips with her. But the point is not over-elaborated, and a change of metre for 'The Good Gossippe's Song' suggests a late addition. There is some realism about the construction of the Ark in this play, and more in the *Pastores*, with its humorous Tudd, and Trowle the *Garcius*, and the rustic banquet. But certainly the *Pastores* has been largely rewritten. The Pilate of the Trial is a more dignified figure than in the later cycles. But the accusers revile the Redeemer as a 'jangellinge Jesu',[1] a 'dosyberde',[2] a 'babelavaunte',[3] and a 'shrewe'. And the buffeters and scourgers do their work with a vigour which is accompanied by a characteristic shortening of the metre.

The York plays are probably of later origin than those at Chester, but how much later we do not know. There are many records of them in the civic Memorandum Book, which begins in 1376. The plays are first mentioned in 1387, but appear then to be well established. In 1394 they were given at stations *antiquitus assignatis*. Their fame invited a visit by Richard II in 1397. Two valuable lists of them compiled by a town clerk, one Roger Burton, are of 1415 and perhaps 1420–2. They come to us in a register written about 1475, and occasionally added to later. Here they are forty-eight in number, but blanks left in the manuscript and various annotations in a late-sixteenth-century hand show that a Marriage at Cana and a Visit to Simon the Leper have been omitted, and that some revised versions, together with *loquela magna et diversa* elsewhere, had never been registered. One of the annotations is addressed to a Doctor, and it may be inferred that they date from 1579 when a revival, the first since 1569, was contemplated, and the council directed that the book should be submitted to the Archbishop and Dean of York for approval. Probably this performance never took place. The council may well have taken alarm at the trouble brought upon Chester mayors by their revivals of 1572 and 1575. Records show that, at one time or another, there were independent plays on the Washing of Feet, the Casting of Lots, the Hanging of Judas, and the Burial of the Virgin, which no longer exist, although fragments of some of them may be incorporated in what we have. Evidently, therefore, the total number of pageants was much greater than at Chester, although individually the York plays are shorter. The second play on the Creation, in fact, consists of no more than a single long speech of eighty-six lines by the Almighty. A good deal of splitting up must have taken place. Moreover, the plays were given on a single day, against the

---

[1] jangellinge, 'talking idly'. [2] dosyberde (dasiberd), 'simpleton'.
[3] babelavaunte, 'babbler'.

three days of Chester. It was certainly a crowded one. The performances began at 4.30 in the morning. The *Peregrini*, which came late, had to be shortened because of the num'ıer of plays still to follow ('for prossesse of plaies that precis' in plight²'), and the Burial of the Virgin was suppressed in 1432, not only on account of the ribaldry which it evoked, but because it could not be given during daylight. It is not surprising that in 1426 the Minorite William Melton, while commending the plays, pointed out that they threw the religious Corpus Christi procession into the background, and suggested that they should be transferred to the following day. It was, however, the procession itself which got so displaced.

Evidently the York texts of the cycle as we have them are not all of one date. Dr. Greg accepts a grouping which involves three periods of literary activity. The first, which he would put about 1350, produced 'a simple didactic cycle, carefully composed in elaborate stanzas and withal rather dull'. This he supposes the work of a single author or a single small school. In a second period it was elaborated by more than one hand. An element of humour was introduced with Noah and the Shepherds, and to this period also belongs 'the work of a writer who is distinguished as being the only great metrist who devoted his talents to the English religious drama as we know it'. The third stage he would also ascribe to a single author, who worked largely on the Passion, and whom he would put not earlier than 1400.

He is a very remarkable though uneven writer. A metrist he certainly is not: he writes in powerful but loose and rugged alliterative verse. He also writes at great length and with much rhetoric and rant. But he is a real dramatist, and his portrait of Pilate is masterly.

I agree largely with this analysis. Certainly the cycle is sharply differentiated from that of Chester by its variety of metrical form. A nucleus of early work is, however, to be found in twelve plays, which are all written in the same stanza of cross-rhyme. It is bipartite, with a four-stressed octave for *frons*, and for *cauda* a three-stressed quatrain on different rhymes. The technical notation is $abababab_4cdcd_3$. This stanza has sometimes been called the 'Northern Septenar', under a misapprehension, since it is not derived from the Latin *septenarius*. It is found in

---

¹ precis (presses). ² plight, 'due order'.

the earliest version of the narrative English *Gospel of Nicodemus*, which was a source for the northern playwrights, and may have been written as early as the first quarter of the fourteenth century. If to the plays composed in this metre we add six in simple quatrains or octaves of cross-rhyme, which are at least as early, we seem to get the outlines of the greater part of an original cycle. It was, as Dr. Greg says, didactic, herein resembling that of Chester, although it had no *Expositor*, and laid less stress upon the element of prophecy. From it survive its Adam and Eve in Eden, Building of the Ark, Abraham and Isaac, *Exodus*, Annunciation, *Pastores*, *Tres Reges*, *Doctores*, Transfiguration, Adulteress and Lazarus, Last Supper, Crucifixion, Harrowing of Hell, Apparition to the Magdalen, Ascension, Pentecost, Assumption, and *Judicium*. I do not know why Dr. Greg ascribes the octaves of the *Judicium* to his metrist. The original Creation has been split up into short plays, and most of it rewritten. If there was a Cain and Abel, it is buried under a late farce. An addition has been made to the *Pastores*. Parts of a Purification survive in a metrical chaos. Most of the Passion has also been rewritten, although a few early fragments remain embedded. There must always have been a Resurrection, but the existing one is of later type than the plays of the nucleus.

For literary quality in the York plays we must look mainly to Dr. Greg's metrist and his dramatist, whom we may also call a realist. As to the latter I have little to add, except that, while he is chiefly to be traced in his revision of the Passion, he also prefixed a new opening to the *Tres Reges*, apparently with the purpose of turning one play into two, and that, while his Conspiracy, Agony, and Betrayal and Condemnation are written in a stanza which may be a development from that employed in the early nucleus, the rest of his work is completely incoherent as regards metrical form. He seems to have picked up and dropped one type of stanza after another, as it suited him at the moment. No doubt the resulting scribal confusion has added to the chaos. Apart from the survival of certain fragments of earlier versions, which he left standing, there is no difficulty in determining the extent of his contributions. They are sharply differentiated from the plays of the nucleus by the combination in them of rhyme with alliterative stress. Herein they fall into line with the alliterative revival in narrative poetry, which developed during the later medieval period, and, although its origin may have been elsewhere, acquired great popularity in the north. The affinity with the narrative writers is very clear in the Condemnation, where the *cauda* is separated from the *frons* of the stanza by the introduction of a one-stress line, known technically as a 'bob'. There is, indeed, a varying amount of alliteration even in the earlier York plays, far more than in those of the Chester cycle. But here it is merely sporadic and ornamental, and the rhythmic movement of the verse remains iambic, with a normal sequence of an unstressed syllable followed by a stressed one. Certainly there are variations. Often the first unstressed syllable is omitted, or an unstressed syllable follows the last stressed one, in what is called a feminine ending. Some lines, as a result, are more naturally read with a trochaic than with an iambic rhythm. Or, again, a stressed syllable may carry two unstressed syllables before it. In alliterative verse this is normal and the number of unstressed syllables may even be greater than two. As a result, the rhythm becomes an anapaestic rather than an iambic one. Alliteration is now used to emphasize the stressed syllables, although it may also occur elsewhere. It often fails, and is sometimes noticeably stronger in the *frons* than in the *cauda* of a stanza.

The determination of the exact extent of the work to be ascribed to the York metrist is a much more difficult problem. Dr. Greg finds him in the Fall of Lucifer and the Death of Christ. Possibly he has also touched that part of the Condemnation which deals with the Remorse of Judas. In these plays he is doubtless a reviser. But I think that we must also give him the *Peregrini*, the Death of the Virgin, and the Assumption, which may not have been handled in the original cycle. These five plays, in different stanza-forms, one of which has a 'bob', are all alliterative. But the alliteration is far less tumultuous than that of the realist, and often falls off in the *caudae* of the stanzas The unstressed syllables, moreover, rarely exceed the limits of an anapaestic rhythm. I am inclined to suggest that a clue to the presence of the metrist may often be traced in a marked tendency to concatenation, the linking up of the beginning of one stanza with the end of that which preceded it by the repetition or slight variation of verbal phrasing. This also is a feature of the non-dramatic poems of the alliterative revival. An outstanding example is the fourteenth-century

Pearl. Isolated concatenations may of course merely arise from the natural give and take of dialogue and carry no significance as evidence of authorship. But it is otherwise when whole poems or continuous groups of stanzas are similarly connected. Concatenation has there become a deliberate literary device. It is so used in the York Condemnation, *Peregrini* and Assumption, but not in the Death of Mary. Our metrist was clearly a versatile writer, and did not tie himself to a single form. Possibly we can go further and find him again in plays which, although in metres other than that of the nucleus, are not, like the five already considered, alliterative. There are four in a stanza-form derived from *rime couée* by dropping two lines of the second *pes*. The technical description is $aaa_4b_2a_4b_2$. The result is markedly lyrical in character, with a dying fall. Often the first four lines are by one speaker, whom another answers in the last two, with something of the effect of the liturgical *versus* and *responsio*. Of these four plays, the Expulsion of Adam and Eve and the Resurrection have a good deal of concatenation, the Incredulity of Thomas a little, and the Temptation none. I think we must ascribe them to the metrist. It is conceivable that he is also responsible for the rather lyrical Adam and Eve in Eden and for the Way to Calvary, both of which are in forms also derived, although differently, from *rime couée*. But there is no clue of concatenation to help us here. And as to the authorship of six remaining plays, I can offer no decided opinion. The Flight to Egypt has an elaborate stanza not paralleled elsewhere in the cycle. Its technical description is $ababcc_4dd_2e_2ff_3$. That of the Fall of Adam and Eve is $abab_4 c_2bc_4dcdc_3$, and that of Joseph's Trouble, the Nativity, the Baptism and the Entry into Jerusalem $abab_4c_2b_4c_2$, with a dying fall, and in Joseph's Trouble and the Entry a tendency to antiphonal speech and reply, which may perhaps again suggest the metrist. The Nativity alone has a little concatenation. Whatever its authorship, there is a good deal of literary merit in this last group. The Entry, with its healing of the blind and lame and its final Hails, is an effective dramatic pageant. And Joseph's Trouble, the Nativity, and the Flight are humane plays, in which the relations between Joseph and the Virgin are touched with a delicate psychology.

From Roger Burton's two lists of the plays and of the companies responsible for them, when compared with some annotations in later hands, with various records in the civic Memorandum Book, and with the state of the register as it has come down to us, it is possible to infer that a good deal of redistribution and probably incidental revision took place between the years 1415 and 1432. We know that in the latter year the Burial of the Virgin was laid aside, and that at the same time the Goldsmiths asked for relief from one of two *Magi* plays which they had hitherto given, on the plea that *mundus alteratus est super ipsos*. This seems to give a likely date for the alliterative opening to the *Magi* written by the realist. His Condemnation may be ascribed to 1423, when an order was made for the combination in a single play of an older version of the theme, together with plays on the Hanging of Judas, the Scourging, and the Casting of Lots, which had apparently been detached from it later than 1415. But there is no Hanging of Judas in the register. Perhaps this episode, in which *Judas se suspendebat et crepuit medius*, had proved too realistic to be witnessed with due sobriety. If then the realist was active between 1415 and 1432, we may perhaps put the metrist in the earlier part of the fifteenth century.

It must be added that, besides the Corpus Christi play, York had another, known as the Creed play. This *ludus incomparabilis* was in the hands of a gild of Corpus Christi, founded in 1408. It had been given by one of its wardens, William Revetor, by a will of 1446, and in 1455 the original manuscript was so worn that it had to be transcribed. It dealt with the articles of the Catholic faith, as set forth in the Apostles' Creed, to which, according to tradition, each apostle had contributed a clause. It was given, instead of the Corpus Christi play, in every tenth year, not on Corpus Christi day, but at the feast of St. Peter ad Vincula on 1 August, which coincided with the harvest festival of Lammas tide. Possibly it consisted of a group of saints' plays. Several performances are recorded during the sixteenth century. But one contemplated in 1568 was abandoned, because Dean Matthew Hutton, to whom the book had been submitted, found things in it which did not agree with the sincerity of the gospel. To it may have belonged the interlude of St. Thomas, the papistical language of which had led to a prohibition by Henry VIII. Of a play on St. Denis of York, bequeathed in 1455 to the church which bore his name, no details are preserved. A *Paternoster* play will be considered later. Evidently the dramatic instinct of the northern city was strong.

The Wakefield cycle, like that of York, has come down to us in a register, written in a hand which may be as late as 1485. There are some gaps, one of which may have contained a Pentecost, or a play on the end of the Virgin. There is no Nativity, which is unusual, but there is no gap at this point. There are thirty-two plays in all, some of which are very long. Two at the end are misplaced additions. A few annotations have been made up to the end of the sixteenth century. In 1814 the manuscript was owned by the Towneley family, of Towneley Hall near Burnley in Lancashire, and the collection was long known as the *Towneley Plays*. It was supposed, according to varying traditions, to have come either from Whalley Abbey near Burnley, or from Wydkirk or Woodkirk Abbey near Wakefield, and it was conjectured that the plays might have been given at a fair held at Widkirk from an early date. There is of course no evidence that a cycle was ever given either in an abbey or at a fair, and none that one existed at Burnley. In the manuscript itself there is much to point to Wakefield as its origin. At the head of it is written in large red letters

In dei nomine amen
Assit Principio, Sancta Maria, Meo. Wakefeld.

This, on the face of it, looks like a title to the whole collection, but there is no separate title for the first play, as there is, again in red, for those which follow. That to the third play runs 'Processus Noe *cum* filiis, Wakefeld'. The others name no locality. Against four of the plays sixteenth-century hands have added 'Barkers', 'Glover Pag', 'Litsters[1] Pagonn', 'lyster[1] play', 'fysher pagent', which at least affords evidence that those plays belonged to a craft cycle. In the *Secunda Pastorum* are two topographical allusions, which might fit either Woodkirk or Wakefield. One is to 'Horbery shrogys',[2] and Horbury is a village two or three miles from Wakefield. The other is to 'the crokyd thorne', which might indeed be anywhere, but might be a 'Shepherd's Thorn' at Mapplewell, near Horbury. So the matter stood when the E.E.T.S. edition of the plays was published in 1897. Since then, further evidence has accumulated, which leaves little doubt that the plays must be

---

[1] litsters (listers), 'dyers'.   [2] shrogys, 'bushes'.

credited to Wakefield itself. The *Mactacio Abel* has the phrase,
'bery me in gudeboure at the quarell hede', and Wakefield had,
as early as the fourteenth century, a lane called Godiboure,
near which was a quarry. I do not accept the inference that
Godiboure is a corruption of 'God i' the bower', and was so
named because the plays were performed there. Other records
have accumulated which show that, in the sixteenth century,
Wakefield had in fact a Corpus Christi play. They begin in
1533 and end in 1566, when the Ecclesiastical Commission laid
down such strict limitations as to what might be shown that
the plays were probably not given. In 1556 they were to be on
Corpus Christi day, but in 1566 at Whitsun. I think we may
now safely regard the plays in the Towneley MS. as a Wakefield
cycle. It must, however, have come into existence at a much
later date than either the Chester or the York cycle. During
the fourteenth century Wakefield was a place of no importance.
A poll-tax return of 1377 shows only one franklin and forty-
seven tradesmen, spread over seventeen occupations, who were
rich enough to be contributory. They cannot have formed gilds
able to maintain plays. There was not much increase by 1395.
But during the fifteenth century the town prospered and became
a head-quarters of the wool trade. A church built in 1329
remained without its tower to about 1409. By 1458 it had been
taken down and rebuilt. One can hardly put the initiation of
a cycle earlier than about 1425. If so, it may have been fairly
complete, on the model of existing cycles elsewhere, from the
beginning. I doubt whether it is worth while trying to analyse
its development into three stages, as at York, although here too,
as at York, there was certainly a comparatively late period of
revision. It is clear that five of the plays have been borrowed,
more or less wholesale, from York itself. They are the *Exodus*,
here called *Pharao*, the *Doctores*, the Harrowing of Hell, the
Resurrection, and the *Judicium*. But they have been differently
treated. The *Exodus* is not much altered. In the other plays
passages have been paraphrased, sometimes in different metres,
and others have been added. The Resurrection has a long
speech for the risen Christ, of which parts are also found in the
Chester Resurrection and in an independent lyric. The *Judicium* has been worked over by a late hand at Wakefield itself.
In the borrowings from York there is much textual corruption.
Dr. Greg does not think that it amounts to more than might

be expected from the normal habits of medieval scribes. Others have suggested that it points to oral transmission, perhaps through the subornation of York actors. But there is nothing to show that certain 'parts' are better represented than others. A reward given at York in 1446 to a *ludens* from Wakefield, which is sometimes cited in this connexion, can of course have no significance. He was probably only a wandering minstrel. The Wakefield debt to York may not have been wholly confined to the five plays already named. A large part of its *Conspiratio* is, like three of those, in the characteristic York stanza of twelve lines, with a four-stress *frons* and a three-stress *cauda*. This may also have been the original form of the York *Conspiratio*, which has been rewritten in an alliterative variant of it by the realist. But I reject altogether a much discussed theory which supposes the York and Wakefield cycles, as we have them, to have gradually developed, through revisions, many of which can now only be conjectured, from a common 'parent cycle'. Apart from the amount of guess-work involved, it is clearly put out of court by a recognition of the fact that the origin of the Wakefield cycle must have been anything from a quarter to half a century later than that of its York predecessor. It is true that parallels of phrasing may often be traced in plays, other than those which Wakefield has admittedly borrowed. They are particularly noticeable in the two plays on Joseph's Trouble. Often they may be due to a common use of narrative sources, such as the *Northern Passion*, the *Gospel of Nicodemus*, or, in the case of Joseph's Trouble, some poem on St. Anne and the Virgin, other than those which have reached us. But occasionally they amount to two or three consecutive lines and, as the Wakefield writers were evidently familiar with the plays at York, they may easily have retained in their memories some noteworthy passages. Whether there was also borrowing from other towns, such as Beverley, we cannot say. The *Prophetae*, with a Sybilla, and scriptural passages annotating the text, is not unlike the manner of Chester.

From the literary point of view, the Wakefield cycle is the best which is preserved to us. The simpler plays use a considerable number of stanza forms, with a predominantly iambic rhythm, and less sporadic alliteration than those of York. And there is a greater tendency to vary the form for different episodes or different speakers within a play. Some of those in

couplets and octaves are rather stiff and dull. On the whole
the *rime couée* and its derivatives predominate. There is not
much use of the cross-rhymed quatrain, either alone or with
a *cauda*, except in the Flight to Egypt, which combines cross
rhyme and *rime couée* in an elaborate stanza with a bob line, of
which the technical description is $ababaabaab_3c_1b_3c_2$. There is
some good poetry in the *rime couée* group, which shows itself
particularly, as at York, in those plays which deal with a human
theme in the varying relations between Joseph and the Virgin.
The Annunciation, in particular, as Dr. Pollard has pointed
out, is 'full of tenderness', and I will quote, after him, the
beautiful verse which describes the occupation of Mary in the
service of the Temple.

> When I all thus had wed hir thare,
> We and my madyns home can[1] fare,
>   That kyngys doghters were;
> All wroght thay sylk to fynd them on,[2]
> Marie wroght purpyll, the oder none
>   Bot othere colers sere.[3]

This touch comes from the *Evangelium Pseudo-Matthaei*, probably
through one of the St. Anne poems. But the outstanding
achievement of the Wakefield plays is to be found in the contributions of a single writer, who is generally known as the
Wakefield Master. There is no tenderness about him, and no
impulse to devotion. He is a realist, even more than his contemporary of York, a satirist with a hard outlook upon a hard
age, in which wrong triumphs over right, but he is saved by an
abundant sense of humour. His contribution consists in the
main of five plays, all written in a characteristic metre of his
own. This too has a 'bob' in it. Its technical description is
$ababab ab_2c_1dddc_2$. The two-stressed lines give it an exceptional
rapidity of movement. There is a good deal of alliteration, but
this does not fall with such regularity on the stressed syllables
as to constitute an alliterative metre, comparable to that of the
York realist. The rhythm, however, is markedly anapaestic,
and may even be called *plus-quam* anapaestic, since a stressed
syllable often carries with it more than two unstressed ones.
The five plays are the Noah, the *Pastores*, a second *Pastores*, the
Magnus Herodes or Innocents, and the Buffeting. In the Noah
the biblical theme of the salvation of mankind through the

---

[1] can (gan), 'began to'.   [2] fynd them on, 'get their living'.   [3] sere, 'various'.

preservation of a single family, who are ultimately to produce the Redeemer, is transformed by this writer into what can only be called a *fabliau* of the recriminations and bouts of fisticuffs which take place between Noah and his wife, from the first building of the Ark to the ultimate return of the dove. The first *Pastores* is much dominated by what may be called a democratic outlook upon the disturbed conditions of the fifteenth century as they affected the rustic working classes. Gyb complains of the 'mekyll vnceyl'[1] of the world. His sheep are rotted; his rent is not ready. John Horne joins him, with laments against boasters and braggers, who do 'mekyll wo' to poor men with their 'long dagers'. Slaw-pace, bringing corn from the mill, enters and chaffs them. Jack *garcio* chaffs them all. They club their poor food together for a meal. Our writer has taken hints from the York play and given them life. Good wine of Ely brings some comfort and the shepherds sleep, after a rustic prayer. An Angel wakes them with news of the Nativity, and bids them go to Bethlehem. They comment, see the star, recall prophecies, and try to sing. The star leads them. And at Bethlehem the tone changes, and the simple gifts of the shepherds, a little spruce coffer, a ball for play, a bottle to drink from, are rather touching. Not content with this effort, our poet essayed another, which is even more audacious. The opening theme is much the same, with complaint of the weather, of the 'gentlery men' who tax the poor, and of the behaviour of wives, especially their own. The world is 'brekyll[2] as glas', and floods drown the fields. Daw's master and the dame bully him, but he will repay them with bad work. The sheep are left in the corn, but the shepherds will sing a part-song. Now enters Mak, a king's yeoman, with a southern tongue. There is mistrust of his honesty. He is a sheep-stealer. His wife drinks and has too many children. And now follows what can only be described as an astonishing parody of the Nativity itself. While the others slumber, Mak steals a sheep and carries it to his wife Gill, who hides it in a cradle, to look like a new-born child. Mak returns to the shepherds, and, when they wake, says that he has dreamed that his wife has had a child, and must go home. The others arrange to count their sheep and to meet at 'the crokyd thorne'. One is missing and they pursue Mak.

---

[1] mekyll (mickle) vnceyl, 'much unhappiness'.
[2] brekyll (brickle), 'brittle'.

There is nothing in his house but empty platters and a cradled child. After a vain search, they go, and return to give the child sixpence. But it proves to have the long snout of a sheep. They can only laugh at the joke, and content themselves with tossing Mak in a canvas. Then comes the angelic message and the visit to Bethlehem, with its gifts. Again the tone has changed.

Primus Pastor

Hayll, comly and clene!
Hayll, yong child!
Hayll, maker, as I meyne,[1]
Of a madyn so mylde!
Thou has waryd,[2] I weyne[3]
The warlo[4] so wylde;
The fals gyler of teyn[5]
Now goys he begylde.
   Lo, he merys;[6]
Lo, he laghys, my swetyng.
A welfare metyng.
I have holden my hetyng;[7]
   Haue a bob[8] of cherys.

Secundus Pastor

Hayll, sufferan sauyoure!
Ffor thou has vs soght:
Hayll, frely foyde[9] and floure
That all thyng has wroght!
Hayll, full of fauoure
That made all of noght!
Hayll! I kneyll and I cowre.
A byrd haue I broght
   To my barne.[10]
Hayll, lytyll tyné mop![11]
Of oure crede thou art crop:[12]
I wold drynk on thy cop,[13]
   Lytyll day starne.

Tertius Pastor

Hayll, derling dere
Full of godhede!
I pray the be nere
When that I haue nede.

[1] meyne (mean).   [2] waryd, 'brought calamity to'.   [3] weyne (ween), 'think'.
[4] warlo (warlock), 'wizard'.   [5] teyn (teen), 'malice'.   [6] merys, 'is merry'.
[7] hetyng (hething), 'scorn'.   [8] bob, 'bunch'.   [9] frely foyde, 'noble creature'.
[10] barne (bairn), 'child'.   [11] mop, 'baby'.   [12] crop, 'completion'.   [13] cop (cup).

Hayll! swete is thy chere!
My hart wold blede
To se the sytt here
In so poore wede,[1]
  With no pennys.
Hayll! put furth thy dall![2]
I bryng the bott a ball:
Haue and play the with all,
  And go to the tenys.

Our poet can lyricize, as well as satirize, when he chooses. But one wonders why Wakefield wanted two *Pastores* plays, and what was actually performed there.

In the *Magnus Herodes* the Master finds himself in his element with the vaunts of Herod and his knights, and in the battle of the mothers to defend their children. And so too in the Buffeting, where the *Tortores* lay on with many insults, as well as blows, and Caiaphas is as violent as Herod and as free-spoken as any Shepherd.

In these five plays the writer is unaided, and there is nothing to suggest that he was a reviser, except of himself. But he also inserted forty-two stanzas of his characteristic metre into the York *Judicium*, to be spoken by Tutivillus and a group of fellow-demons, in scarification of the unjust souls, whose records they bring. Herein is much social criticism, of fraudulent tax-gatherers, perjurers, extortioners, simoners, lechers, and adulterers, and of the extravagance in dress of men and women. But it is Tutivillus who is a 'mastar lollar'.[3] The Wakefield writer was no heretic. An attempt has been made to date his work as not earlier than about 1426, from the references to costume. I do not find the evidence very convincing. Sartorial fashions rise and fall and rise again. One feature, in particular, which is relied upon, that of the woman's head-dress 'hornyd like a kowe', was already the subject of unkindly comment by John de Bromyard, as far back as 1360. Some traces of east Midland influence on the Master's northern dialect also seem to point to about 1426. But we have seen that the inception of the Wakefield cycle as a whole may not have been much earlier than this. The Master may therefore have been one of its original writers. I doubt whether he was influenced by the York realist, who writes, not merely with an anapaestic rhythm,

---

[1] wede (weed), 'clothing'.   [2] dall, 'hand'.   [3] lollar (lollard).

but with the full alliterative stress. They may have been contemporaries, and perhaps rivals. Both, no doubt, use bob-lines, but the Master's two-stressed *frons* does not appear at York. Several of the Wakefield plays, the Conspiracy, the Scourging, the Crucifixion, the *Processus Talentorum*, which should be *Talorum* (dice), the *Peregrini*, the Ascension, the Lazarus, are again medleys, and have probably been revised. A few of the Master's stanzas are in the Conspiracy, the Scourging, and the *Processus Talentorum*. There are bob-line stanzas elsewhere, but they lack his two-stressed *frons*. Nor do I think that he has any responsibility for the *Mactacio Abel*, into which a farce, more elaborate than that of York, has been inserted.

From the Coventry cycle only two plays have come down to us, and those in very late and degraded forms. Local annals, of which the earliest are themselves only in a manuscript of about 1587, put the initiation of the cycle in 1416. There were certainly pageants earlier. The Drapers had one by 1393 and the Pinners by 1414. It is possible that these were only dumbshows. A gild of Corpus Christi was founded in 1348, and was later merged in the corporation. In the ridings, *ex antiquo tempore*, from seventeen to twenty-two crafts took part about the middle of the fifteenth century. But for dramatic purposes these were combined into ten groups, of which the most important members took financial responsibility for the production of the plays, while the others were contributory. The ten plays appear to have been given at ten stations, one in each of the ten wards of the city. They were long, and crowded with characters, and some of the action apparently took place on supplementary scaffolds and even in the street. The two extant plays are those of the Shearmen and Tailors, and the Weavers. In the first a prophetic prologue by Isaiah is followed by a series of distinct episodes, covering the Annunciation, Joseph's Trouble, the Journey to Bethlehem, and the *Pastores*. A dialogue by two unnamed Prophets separates these from a second series, which has Herod and the *Magi*, the *Magi* at Bethlehem, a brief Flight to Egypt, and an Innocents. It has been suggested that two original plays are here combined, but there is no evidence that the Shearmen and Tailors ever formed separate gilds. The Weavers' play is similarly constructed, with a long prophetic dialogue for opening, followed by a Presentation in the Temple, a Visit of Joseph and Mary to Jerusalem, and a

Doctores. The *Doctores* seems to be a late addition, with a text derived, as at Chester, from one like that of York, through some intermediate form. It may have been borrowed in 1520, when the annals record that there were 'new playes at Corpus Christi tide'. Of the other eight plays we can trace, mainly from gild accounts, a Trial and Crucifixion, a Burial, a Harrowing of Hell, with a Resurrection and *Peregrini*, and a *Judicium*. The rest must remain uncertain. There is some slight evidence for an Assumption. A Baptism, Entry into Jerusalem, Last Supper, Ascension, and Pentecost are all possible. It seems clear that there were no Old Testament plays at Coventry. The Prophets, of course, had long been attached to the Nativity theme. During the latter part of the fifteenth and the sixteenth centuries the Coventry plays seem to have had a more than local reputation. The Midland town was more accessible from London than those of the north. Several royal visits are upon record. A preacher on the Creed in 1526 bids his audience go to Coventry and see what he has told them represented there. John Heywood, in his *Foure P.P.* of about 1545, refers to one who had 'playd the deuyll at Coventry'. Certainly the revision of 1520 was not the last. According to the annals, the plays were given in 1568, after which they were 'laid down' for eight years, then revived in 1576, and last played in 1580. The texts which we have were 'correcte' or 'translate' by one Robert Croo in 1535, and to them are appended some rather pretty Elizabethan songs by Thomas Mawdycke, dated as late as 1591. This double recension has unfortunately left the plays completely unreadable. Much is a mere jumble of incongruous metres. One can at most discern in the chaos an original basis, as elsewhere, of quatrains, octaves, and *rime couée*, and a probably later addition of rhyme royal and its four-stressed equivalent. But all has been turned, by the clumsy insertion of superfluous unstressed syllables, which destroy the iambic rhythm without substituting an anapaestic one, into what can only be described as doggerel. It has sometimes been called 'tumbling verse', a term for which James VI of Scotland seems to be responsible. But I do not know that that is any more polite. There is also much aureate language, and in particular a constant use of latinized words ending in -*acion*, with a stress falling on the antepenultimate syllable. Apart from the cycle, the annals record saints' plays on St. Katharine in 1490, and apparently St. Christina in 1504, which were given in the Little Park and may have been stationary performances.

Outside the four craft-cycles already described, a few other plays of the same type are preserved. Two of these are certainly also craft-plays. From Newcastle-on-Tyne comes a Noah. Here the plays appear to have been stationary, in a 'stead', after the Corpus Christi procession was over. They are first recorded in 1427, and the extant text may be of much the same date. It only deals with the Making of the Ark, and probably a Deluge followed. The dialogue is mainly in quatrains, with occasional variants in other metres. It is fairly lively. A *Diabolus* gives Noah's wife a drink for her husband, presumably to check his enterprise, but so far the farcical element is not as strong as in the York and Wakefield versions of the theme. Of a Norwich Creation of Eve and Expulsion from Eden there are two late texts, one of a fragment of about 1533, the other, complete, of 1565. Both are in the seven line four-stressed stanza, with much doggerel and some aureate language. Among the characters are two allegorical figures, Dolor and Myserye. The records of the cycle go back to 1527, when the gild of St. Luke handed it over to the crafts. Possibly it existed as early as 1478, when a letter to Sir John Paston, written from Norwich on May 20, compared the behaviour of the Earl of Suffolk to one playing Herod in Corpus Christi play. The sixteenth-century performances at Norwich, however, were not on Corpus Christi day but in Whitsun week. An Abraham and Isaac, found in a manuscript at Trinity College, Dublin, probably only got there by accident. There is a record in 1498 of a dumb-show procession on Corpus Christi day at Dublin, but none of plays. The same manuscript contains a list of Northampton mayors up to 1458, and possibly the play came from that town, although there is no evidence that it ever had a cycle, beyond a rather vague reference to the storing of pageants in a hall there during the half-century before 1581. The dialect, however, seems to belong to the border between the east and west Midlands. The play, written mainly in *rime couée*, is not a bad one. The introduction of Sara as a character helps to humanize its theme. It has been conjectured that the writer knew some version of the French *Viel Testament*. A second Abraham and Isaac comes from a manuscript of 1470–80 preserved at Brome Manor in Norfolk. It is dull, and mainly

in doggerel. There is nothing to show whether it formed part
of a cycle. Opinions have differed as to its relation to other
dramatic versions of the story. Dr. Greg, who has made a
minute study of their variants, thinks that the Chester one is
the earliest extant, and that the Brome one may be derived
either from that or from a common original. A play on the
Innocents and the Purification is found in Bodleian Digby MS.
133, which is a composite manuscript in more than one hand,
and apparently contains pieces from different sources. This
one has been glossed by a later hand with the date 1512 and
the statement 'Iohn Parfre ded wryte thys booke'. Yet another
hand, that of the chronicler John Stow, has added 'the vij
booke'. An opening speech by a Poet shows that the play
belonged to a cycle, but one spread over a series of successive
years, and performed, like the Lincoln plays, on St. Anne's day.
A *Pastores* and a *Reges* had come in the year before. There is
nothing to suggest that the cycle was undertaken by craft gilds.
Again we have unfortunately little but doggerel. The dialect
points to a Midland author and a southern scribe. Finally, a
very curious *Burial and Resurrection*, in a mid-fifteenth-century
hand, is in Bodleian MS. E. Museo, 160. I do not know why
Furnivall thought that it was once part of the Digby MS. It
was originally written as a 'treyte or meditatione' and then
converted into a play to be given in part on Good Friday
afternoon, and in part on Easter Day, thus becoming an isolated
vernacular example of a liturgical play, extending to 1,631 lines.
The Latin dialogue *Dic nobis, Maria* is incorporated, and there
are directions for the use of the sequences *Victimae Paschali* and
Scimus Christum. There is no evidence and little probability that
the piece was ever performed. The dialect is northern, with
some Midland forms. The metre is mostly *rime couée*, with some
element of doggerel. There is aureate language. An elaborate
lament of the Virgin is in octaves, and makes frequent repetition
of the line 'Who cannot wepe, com lerne at me', which is also
found as the burden of two fifteenth-century lyrics.

The Digby MS. also contains two saints' plays. In view of
the number of such representations disclosed by records, it is
surprising that so few have survived. A *Conversion of St. Paul*,
in four-stressed seven-line stanzas, is doggerel with aureate
touches. The dialect is east Midland. A second hand has
inserted a scene between two devils, Belial and Mercury. And

against a prologue, headed *Poeta*, another has written 'Myles Blomefylde'. An alchemist of that name was born at Bury St. Edmunds in 1525. The performance was given at three 'stations', apparently spots in an open space, to which the audience moved successively. More interesting is a *St. Mary Magdalen*, in the second hand of the *St. Paul*, with the initials M. B. at the beginning. This too is in Midland dialect. The versification is much the same as that of the companion play, but even more metrically incompetent, although not without some rough vigour. The whole life of the saint, as related in the *Legenda Aurea*, is covered, with action at a number of spots, for which elaborate stage-directions are given. Between them, a practicable ship comes and goes. 'I desyer the redars to be my frynd', says the writer at the end, and perhaps he hardly expected the play to be performed. To one of its episodes it will be necessary to return later. Yet another isolated text requires consideration. It is not hagiological, but it is a *miraculum* in the strictest sense, being based upon an actual miracle, which according to an appended note, signed, presumably by the author, R. C., took place at Eraclea, in the forest of Aragon, during 1461. The text, entitled *The Conversyon of Ser Jonathas the Jewe by Myracle of the Blyssed Sacrament*, may be not much later. But in fact the theme is traceable in continental legend, and even on the stage, as far back as the fourteenth century. From a literary point of view this is perhaps the most interesting dramatic relic, outside the craft cycles, which has come down to us. A Banns is prefixed, in which two *vexillatores* announce a performance at Croxton. There are several places so named, but as the dialect is east Midland, either one in Norfolk or one in Cambridgeshire seems most likely. A character is said to live near Babwelle Mill, and at Babwell, two miles from Bury St. Edmunds in Suffolk, was a Franciscan priory. Possibly the play originated at Bury, and was taken on tour, as we know that a Chelmsford play was in the sixteenth century. The action of the *Sacrament* opens with a long speech by a merchant Aristorius, who vaunts the range of his traffic, and sends his clerk to learn if any ships have come in

Of Surrey[1] or of Sabe[2] or of Shelys-down.

There may be a topical reference here. A Shelley is near the

[1] Surrey, 'Syria'.

[2] Sabe, either Sheba, or Saba, an island in the Dutch West Indies.

estuary of the Stour in Suffolk, but I do not know whether it
ever had a harbour. The passage recalls one in the Wakefield
Magnus Herodes, where the Nuncius similarly vaunts his lord's
renown in many lands,

> From Egyp to Mantua
> Unto Kemp towne.

Another character in the *Sacrament* is a Jewish Jonathan, who
wants to test the truth of the Christian story of a cake, which a
priest can turn into the flesh and blood of him 'that deyed upon
the rode'. He approaches Aristorius, who is friendly with the
priest Ser Ysodyr, and bribes him to dope the priest with wine,
and steal a Host from the church, while he sleeps it off. The
plot succeeds. Jonathan summons four other Jews. They prick
the Host with their daggers, and it bleeds. They bring a
cauldron of oil to boil it in, but it clings to Jonathan's hand,
and he runs mad. The others nail him to a post, and pull the
Host away, but his hand comes with it. The aid of a leech,
Master Brundyche of Braban, is suggested, and an elaborate
proclamation advertises his skill. He recalls the quack doctor
of the folk-plays, who may also have influenced the *unguentarius*
of some versions of the *Visitatio Sepulchri*. But he is sent away as
useless. The Host is then thrown into the cauldron, the oil in
which turns into blood. Finally it is baked in an oven, which is
riven asunder, and Christ appears, as an image with wounds
bleeding, and makes an appeal to the Jews. They repent, and
at the prayer of a bishop the image becomes bread again.
Aristorius in his turn repents. The rather odd moral is drawn,
among others, that curates ought to keep their pyxes locked.
It was an audacious piece to put before our forefathers in
an age of devotion. But the author has a vigour of expression, unmatched in medieval drama, except by the Wakefield
Master. He uses various stanza-forms, and can hold either an
iambic or an anapaestic rhythm at will. In some passages
only he alliterates his stressed syllables. Occasionally he uses
an aureate phrase, such as 'mansuete[1] myrth' or 'largyfluent[2]
Lord'.

The evidence of these scattered plays suggests that a considerable variety in the manner of dramatic representation

---

[1] mansuete, 'gentle'.   [2] largyfluent, 'copious'.

prevailed in the east Midland area. And this is perhaps confirmed by one other important text from that area, which has
still to be discussed. This is the so-called *Ludus Coventriae*, about
which almost the only certain thing is that it has nothing to do
with Coventry. The misapprehension may have started with
Robert Hegge, the first known owner of the manuscript, who
was a Fellow of Corpus Christi College, Oxford, and died in
1629. It was taken up by his fellow-collegian, Richard James,
librarian to Sir Robert Cotton, who wrote *Ludus Coventriae sive
Ludus Corporis Christi* on a fly-leaf of the manuscript, and by
Sir William Dugdale in his *Antiquities of Warwickshire* (1656),
who alleged that the plays were given by the Grey Friars of
Coventry. This probably rests on a statement in a late version
of the Coventry *Annals*, which records that in 1492–3 Henry VII
saw the plays 'by the Gray Friers'. These were doubtless the
craft-plays, which the king presumably watched from a station
at the gate of the Grey Friars convent, and the existence of a
second cycle in Coventry is as unlikely as the performance of
any cycle by a religious house. The dialect of the *Ludus* seems
to be an East Anglian variety of east Midland. To the text of
the plays is prefixed a Banns by three *vexillatores*, who enumerate
the 'pagents' and their subjects, refer to 'oure pleyn place', and
announce a performance to begin on a following Sunday at
6 a.m. 'in N. towne'. It was evidently to be a stationary performance, and there is no mention of crafts. 'N. towne' might
of course fit either Norwich, which, however, had craft-plays,
or Northampton, where we have no clear record of plays. But
I suspect that N. merely stands, as it does in the Marriage
Service, for *Nomen*, and that the *Ludus*, like the *Sacrament*, was a
touring play. The Banns contemplate a complete cycle of
forty plays. The text, which in many respects does not
agree with the Banns, has forty-two. But both documents have
undergone revision. The metre of the Banns is a thirteen-line
stanza, with a *frons* and *cauda*, of which the technical description
is $ababab ab_4 c_3 ddd_4 c_3$. This is also the metre of some of the plays,
but others are in *rime couée*, or in octaves, and occasionally
quatrains or couplets. Dr. Greg, who has made a valuable
study of the manuscript, thinks that some of these have been
introduced by successive revisers from other cycles, but for this
I do not see any very strong reason, in view of the variety of
metrical form shown in the York and Wakefield plays. What

seems clearest is that the whole of the cycle, and with it to some extent the Banns, has been worked over by a late writer, who has added much material of his own composition, again in octaves, which can, however, be easily distinguished from those of the original by their long doggerel and aureate lines. According to Dr. Greg, he handed instalments of his revision to a scribe, who contributed further corruption to the text as we have it. The date 1468, which is found on one page of the manuscript, may well be that of his work. The activity of this writer is, I think, the main factor in the state of confusion to which the cycle has been reduced. The text, as it stands, does not agree with the Banns, or with the method of performance which they imply. The reviser, who also uses the East Anglian dialect, was a theologian, with a desire to expound at length the significance of moral and liturgical formulas, such as the Ten Commandments, the Psalter, or the Steps to the Temple. He had a special devotion to the 'clene mayd', the Virgin Mary, and he has added three plays on her Conception, her Presentation in the Temple, and her Visit to Elizabeth, of which there is no trace whatever in the Banns, and for which he seems to have borrowed some material from one of the extant poems on St. Anne. Moreover, he has arranged them in a way which is quite inconsistent with the Banns. Together with the Betrothal and Marriage, the Annunciation, and the Trouble and Return of Joseph, they make a self-contained group, which is introduced by a presenter *Contemplatio*, who informs the 'congregacion' that the matter of the 'processe' is of 'the modyr of mercy', makes further comment at intervals, and at the end thanks the audience for their patience, and adds,

With Avé we begunne and Avé is our conclusyon.

Clearly the reviser has departed from the scheme of the Banns, and has arranged these six plays as a self-contained performance. Later comes a Purification, for which also the Banns, as we have them, do not provide, but which is not, I think, the work of the main reviser. He has, however, wholly recast and largely overwritten the Passion plays, inserting, rather oddly before the Entry into Jerusalem, a long prefatory speech by a Demon, a prophecy by John the Baptist, and a Conspiracy of his own. And here again, he contemplates a method of representation which is quite other than that of the Banns, for between the

Captio and the Trial before Herod appears an *Expositor*, who is again *Contemplatio*, and announces to the audience,

We intendyn to procede the matere that we lefte the last ȝere.

And here, I think, his main work ended. The later plays are mostly in their original metres, and he can only have lightly touched them. An exception is the Assumption, written in octaves and thirteen-line stanzas, curiously linked by inter-calary lines. It is on an interpolated quire, in a hand other than that of the usual scribe. Of this also the Banns know nothing. Dr. Greg would, however, ascribe it to a writer distinct from the main reviser. I should add that at some time the main reviser appears to have made an attempt to alter the Banns themselves, in order to bring them into conformity with his expanded cycle. But this enterprise he must have dropped half-way. Obviously the revised *Ludus* can never have been performed as a whole. Dr. Greg points out that the text, ornamented with elaborate genealogical tables and notes on the dimensions of the Ark, was evidently intended for readers. Probably the reviser was a member of some religious house, which had a library. One thinks of Lincoln, where there was a special cult of St. Anne, and where a *Ludus Corporis Christi* was occasionally played in the latter part of the fifteenth cen-tury, or of Bury St. Edmunds. But the main reviser was not Lydgate, as has been suggested. When he ends a line, as he so often does, with a word terminating in -*acion*, he puts his last stress on the antepenultimate, but Lydgate, in a like case, on the ultimate syllable. As literature, of course, much of the *Ludus Coventriae* in its present state makes difficult reading. But some of the unaltered plays are not without lyrical merit, notably the *Pastores*, the *Magi*, and the Resurrection. Some passages of short-lined *rime couée*, apparently regarded as appropriate to *Milites*, are particularly attractive.

The Miracles were not the only form of dramatic expression available for the instruction and incidentally the entertainment of our medieval ancestors. Somewhat later, but still before the end of the fourteenth century, emerge the Moralities. Here the subject-matter is still religious, but the angle of approach is different. The Corpus Christi play was based entirely upon Holy Writ, with certain legendary material which had attached itself to the divine record in the course of ages. It had its one great theme of the Redemption. In terms of dramatic art, it was a divine comedy rather than a tragedy. There was a strong element of pathos in the Passion, but this was followed by the triumph of the Resurrection. There was an underlying didactic intention, no doubt, but this remained subordinate, in such episodes as that of Moses and the Tables, or the *Doctores*, until a final moral was drawn in the epilogue of the *Judicium*. The morality makes a fresh start, just where the miracle play left off. It is wholly concerned with the behaviour and final destiny of man. And if the miracle play derives from the liturgy, so does the morality from the homily, which had become a prominent feature in English religious life, with the revival of preaching by the Dominican and Franciscan orders in the twelfth and thirteenth centuries. Dr. Owst has rightly dwelt upon the close analogy between the topics and manner of the playwrights and those of the contemporary pulpit. The method adopted was largely that of allegory. It had already been used by early Christian writers. During the fourth century Prudentius had represented the conflict of spiritual forces for the soul of man as a siege in his *Hamartigenia* and as a battle in his *Psychomathia*. From him comes the conception of the Seven Deadly Sins, which came to be generally regarded as Pride, Lust, Sloth, Envy, Anger, Avarice, and Gluttony, and to which were opposed the Seven Cardinal Virtues, Faith, Hope, Charity, Justice, Prudence, Temperance, and Fortitude. Bernard of Clairvaux and Hugo of St. Victor, in the twelfth century, had taken a passage from the Psalms, 'Mercy and truth are met together; righteousness and peace have kissed each other', and turned it into a Debate of the Four Daughters of God, in which Mercy and Peace plead in heaven against Truth and Righteousness for forgiveness to the Soul of Man. On these and similar hints the writers of the moralities devised a type of performance in which the characters were no longer, as in the miracles, scriptural personages, but sheer abstractions, and the action took in one way or another the form of a struggle of Good and Evil for the possession of Man, who is himself, under one name or another, the central figure. It sounds rather a bloodless affair, but it must be remembered that on the stage the participants looked much more like ordinary men and women than they can in cold print. The playwrights, moreover, showed considerable ingenuity in varying the presentation of the theme and providing picturesque detail. In the miracle plays themselves abstract personages rarely appear, outside the *Ludus Coventriae*, which is late enough in its present form to have been a borrower from the moralities. Besides the presenter *Contemplatio* himself, it has a version of the Four Daughters of God in the Conception of the Virgin. That is by the main reviser, but *Mors* may possibly have come for Herod in the Innocents, before that reviser touched the play. The Mary Magdalen play of the Digby MS. has been similarly affected.

The earliest morality of which we know anything is a *Pater Noster* play. A tract in favour of an English translation of the Bible refers to friars who had taught 'the paternoster in engliȝsch tunge, as men seyen in the pley of York'. If it is rightly ascribed to Wyclif, it must have been written before his death in 1384. In any case a Gild of the Lord's Prayer existed at York in 1389, and a return then made shows that it had a 'ludus de utilitate Oracionis Dominice compositus', in which 'quam plura vicia et peccata reprobantur et virtutes commendantur'. The *sapor* of this play had led to the formation of the gild, whose brothers and sisters were bound to maintain it, and to walk with it in the streets, on what day is not said. A *compotus* of 1399 shows that it included a *ludus Accidie*, which is Sloth. Later, apparently by 1462, the play had passed into the hands of the Mercers' craft gild, who then or in 1488 seem to have produced it at Corpus Christi, possibly as a sequel to their *Judicium* in the cycle. In 1558, when the book was held by the Master of St. Anthony's, apparently a hospital which had escaped suppression, another performance took place at Corpus Christi, at the expense of the crafts. A final one came in 1572, after which the Archbishop of York sent for the book, which he never returned. York was not the only place which had a *Paternoster* play. One was given by the crafts of Beverley in 1469, not at Corpus Christi, but on the Sunday after the feast of St. Peter ad Vincula in August. There were seven stations and eight pageants, of *Superbia*, *Luxuria*, *Accidia*, *Gula*, *Invidia*, *Avaricia*, *Ira*, and *Vitiosa*. Local annals also record, without detail, performances of a *Paternoster* at Lincoln in 1398, 1411, 1425, 1456, and 1521. A connexion between the Lord's Prayer and the Seven Deadly Sins is intelligible enough. The prayer can be read as containing seven petitions. St. Augustine had treated them as calling for the sevenfold operation of the

Holy Ghost, described in the Vulgate version of the Book of Isaiah as 'the spirit of wisdom and understanding, the spirit of counsel and might, the spirit of knowledge and piety, and the spirit of the fear of the Lord'. It seems again to have been Hugo of St. Victor who linked this enumeration with that of the seven deadly sins, and the seven virtues opposed to them. From him, too, we can infer an explanation of the eighth pageant at Beverley, which was that of *Vitiosa*. This figure must have stood for frail Humanity itself, not necessarily sinful, but with tendencies impelling to sin. 'Hoc autem interesse videtur inter peccata et vitia', says Hugo, 'quod vitia sunt corruptiones animae ex quibus, si ratione non refrenentur, peccata, id est actus injustitiae, oriuntur.'

Another theme, hardly less prominent in the moralities than the Conflict of Vices and Virtues, is that of 'Dethe, Goddys mesangere'. The fleetingness of life and its glories has of course been a preoccupation of thinking men from the earliest ages. In English literature it may be traced as far back as the *Wanderer* and the *Metres of Boethius*. It gave rise to the long series of *Ubi sunt* poems, and to the *Sayings of St. Bernard* and the *Love Rune* of Thomas of Hales in the thirteenth century.

> Hwer is Paris & Heleyne
>   That weren so bryht & feyre on bleo,[1]
> Amadas & Ideyne,
>   Tristram, Yseude and allé theo,
> Ector, with his scharpé meyne,[2]
>   & Cesar, riche of wordés feo?[3]
> Heo beoth i-glyden[4] vt of the reyne[5]
> So the schef[6] is of the cleo.[7]

The imminence of death could not be overlooked in the pestilences of the fourteenth century or the anarchical conditions of the fifteenth. It was the constant theme, in its religious aspect, of the pulpit. In a set of Latin verses, known as the *Vado Mori* poem, twelve representatives of all estates, from Pope to Pauper, speak appropriate couplets. That of the King runs,

> Vado mori, rex sum: quid honor, quid gloria regum?
>   Est via mors hominis regia. Vado mori.

---

[1] bleo (blee), 'aspect'.   [2] meyne (main), 'strength'.
[3] wordés feo, 'world's wealth'.   [4] i-glyden (glided).   [5] reyne, 'realm'.
[6] schef (sheaf?).   [7] cleo (cliff), 'hill-side'?

This may be the precursor of the lines known in France as the *Danse Macabre*, and in England as the Dance of Death. It has been conjectured that they may have owed their origin to a quasi-dramatic ceremony, in which a priest pronounced them from the pulpit, while appropriately clad figures, led by Death, passed to a tomb in the nave of the church. But they are only known as attached to painted mural representations of such figures, of which the earliest, at Klingenthal in Saxony, is of 1312. There was a famous one in the church of the Holy Innocents at Paris, from which Lydgate adapted verses for another, set up about 1430, in a cloister round Pardon Church-yard near St. Paul's, which was pulled down to provide building material for Somerset House in 1549. There were others at Salisbury Cathedral, Stratford-on-Avon, and elsewhere in England. Some verses preserved in manuscripts may come from one or other of them:

> I wende[1] to dede[2] knight stithe[3] in stoure,[4]
> Thurghe fyght in felde I wane the flour,
> Na fightis me taght the dede to quell,
> I weend to dede, soth[5] I ʒow tell.

Dramatic performances based on the Dance of Death are recorded at Bruges in 1449 and at Besançon in 1453.

In England, apart from the incursion of *Mors* into the Innocents of the *Ludus Coventriae*, there is little trace of the influence of the Dance upon the miracle plays. Here the end of man is in the *Judicium*. It is otherwise in the moralities, and particularly in the earliest of which a text has come to us. It is known as *The Pride of Life*, although *The King of Life* would be a better title. The manuscript is written on the back of an account roll by two northern scribes, but the rhymes point to a southern author, perhaps in the first decade of the fifteenth century. Unfortunately only fragments of the play are preserved. There is a long prologue by a presenter, who foreshadows the action. It is a 'game' in an open 'place'.

> Nou stondit still & beth hende,[6]
> And prayith al for the weder.

There is a tent for the king and a 'se' for a bishop. *Rex vivus*

---

[1] wende, 'go'.   [2] dede (death).   [3] stithe, 'stout'.
[4] stoure, 'fight'.   [5] soth, 'truth'.   [6] hende, 'courteous'.

is supported by two soldiers, who are Strent and Hele in the
text and *Fortitudo* and *Sanitas* in the rubrics. He vaunts,

> Qwher of schuld I drede
> Qwhen I am king of life?
> Full evil schuld he spede,
> To me that worth[1] striue.
>
> I schal lyue ever mo
> & croun ber as kinge;
> I ne may neuer wit of wo;
> I lyue at my likinge.

To him comes his queen and bids him remember his latter
end, but he rejects the counsel. Mirth, also called Solas, his
Nuncius, flatters him, and is promised the castle of Gailispire on
the hill and the earldom of Kent, which in fact became vacant
in 1407. The king retires to his tent, and the queen sends Mirth
for a bishop to exhort him. The bishop laments the state of the
world in terms which recall the *Pastores* of the Wakefield plays.
The rather eccentric spelling of the text is here normalized.

> The world is now, so wo lo wo
> In such bale ibound,
> That dread of God is all ago,
> And truth is gone to ground.
>
> Peradventure men hold me a fool,
> To sing that fool tale;
> They fare as fishes in a pool:
> The great eateth the small.
>
> Rich men spare for nothing,
> To do the poor wrong:
> They thinketh not on their ending,
> Nor on death, that is so strong.

He preaches to the king, who calls him a 'bissop-babler' and
a 'chagler',[2] and sends Mirth abroad to cry the 'banis'[3] of his
defiance of Death. And there the fragments break off, which
is a pity, for the writing is rather forceful in its sombre way, in
spite of some awkwardness in the handling of the metre. The
outcome can, however, be gathered from the prologue. Death
appeared and brought a dreadful dream, of father and mother

---

[1] worth (worketh).   [2] chagler (cackler).
[3] banis (banns), 'proclamation'.

and the king himself slain. Then followed a stern strife between
Life and Death, 'the king of life to wrake'.[1] The Fiend took the
soul. But at the end comes a note of reconciliation,

> And oure lady schall ther for preye
> So that with her he schal be lafte.

A far more elaborate piece is *The Castle of Perseverance*, which
runs to 3,650 lines and has thirty-four characters. It is one of
three independent plays in a collection known as the Macro
MS., the history of which suggests that it may have come from
the Abbey of Bury St. Edmunds. The scribal hand is of about
1440, and the date of composition probably not much later
than about 1425. The dialect is that of Norfolk or Lincolnshire,
and there is a mention of the 'galows of Canwyke', which has
been traced near Lincoln. The manuscript contains an elaborate
plan for the setting of the performance. A circular place is to
be kept free of spectators, and surrounded, if possible, by a
ditch filled with water, but, if not, barred about. There must
not be too many sticklers to prevent a good view. The castle is
to be in the centre, and beneath it a bed from which a soul can
arise, and a cupboard at the foot of the bed. It is not clear at
what point of the play this was to be used. Outside the ring
must be five scaffolds, for *Deus*, *Mundus*, *Caro*, Belial, and
Covetyse. The characters generally get English names in the
text but Latin equivalents in the rubrics. The action centres
about the career and final destiny of *Humanum Genus* or Mankind. A prologue by two banner-bearers gives an outline of the
plot, and promises a performance in a week's time, at a place
the name of which is left blank. It will begin at undern, which
is 9 a.m. The prologue does not quite square with the play itself
and may be a later addition. It anticipates a conclusion by grace
of 'oure lofly lady', the Virgin, who does not in fact appear.
The plot is a complicated one. *Mundus*, Belial, also called
Satanas, and *Caro* boast on their scaffolds. Says *Caro*,

> Be-hold the Werld, the Deuyl, & Me!
> With all our mythis,[2] we kyngys thre,
> Nyth[3] & day, besy we be,
>> for to distroy Mankende,
>> If that we may.

---

[1] wrake (wreak), 'harm'.   [2] mythis (mights).   [3] nyth (night).

Mankind describes the feeble state in which he was brought into life 'to woo & wepynge', and his Good and Bad Angels. They are present and discourse with him. He decides to go with his Bad Angel to *Mundus* and get riches. *Mundus* sends Folly and Lust to fetch him, and he proclaims his choice:—

> Mary, felaw, gramercy!
> I wolde be ryche & of gret renoun.
> Of God I yeue no tale[1] trewly,
> So that I be lord of toure & toun,
>   Be buskys[2] & bankys broun.
> Syn that thou wylt makë me
> Bothë ryche of gold & fee,
> Goo forthe! for I wyl folow thee
>   Be dale & euery towne.

He pledges his truth to *Mundus*, and Folly and Lust become his servants. Backbiter (*Detractio*), the messenger of *Mundus*, will teach him the way to the Seven Deadly Sins, Covetyse, the treasurer of *Mundus*, gives him wealth and introduces him to the other six. Pride, Wrath, and Envy come from the scaffold of Belial, and Gluttony, Lechery, and Sloth from that of *Caro*. There is some free-speaking on the details of sin. Mankind has 'mekyl myrthe', but expects an end in hell. The Good Angel laments; the Bad Angel rejoices. But now Shrift, apparently called both *Penitentia* and *Confessio* in the rubrics, comforts the Good Angel and leads Mankind to repentance. He must, however, enter the Castle of Perseverance for security. A leaf is missing here, but he is presumably welcomed by Patience, as well as by the other six virtues, the ladies Charity, Abstinence, Chastity, Solicitude, Generosity, and Humility. But all is far from over. The spiritual life calls for endurance as well as repentance. The Bad Angel is ready for action.

> Nay! be Belyals bryth bonys[3]
>   Ther schal he no whyle dwelle.
> He schal be wonne fro these wonys,[4]
>   With the Werld, the Flesch, & the Deuyl of hell:
>   Thei schul my wyl a-wreke.[5]
>   The synnys seuene, the kyngis thre,
>   To Mankynde haue enmyté;
>   Scharpely thei schul helpyn me,
>   This Castel for to breke.

---

[1] yeue no tale, 'take no account'.   [2] buskys (bushes).   [3] bryth bonys (bright bones).
[4] wonys (wones), 'abodes'.   [5] a-wreke (wreak), 'gratify'.

Backbiter is sent for *Mundus*, Belial, and *Caro*, who thrash the Seven Sins for letting Mankind get into the castle. An attack upon it is ordered. Mankind prays to 'that dynge[1] duke that deyed on rode' for grace to resist temptation. The Good Angel calls on the Virtues, and defiances are exchanged between them and the Vices. *Tunc pugnabunt domini*, says the rubric. The troops of Belial and *Caro* are beaten, with the roses flung by their adversaries. It is a pretty touch. But the resource of *Mundus* is not exhausted. He sends once more Covetyse, who finds Mankind in poor estate.

> Couetyse! whedyr schuld I wende?
> What wey woldyst that I sulde holde?
> To what place woldyst thou me sende?
> I gynne to waxyn hory & olde;
> My bake gynnyth to bowe & bende;
> I crulle[2] & crepe & wax al colde;
> Age makyth man ful vnthende,[3]
> Body & bonys, & al vnwolde.[4]
>   My bonys are febyl & sore.
> I am arayed in a sloppe;[5]
> As a yonge man, I may not hoppe;[6]
> My nose is colde, & gynnyth to droppe;
> Myn her[7] waxit al hore.[8]

He is an easy prey. The Virtues lament, but Mankind leaves the castle and goes with Covetyse, who gives him a thousand marks, with the strict injunction that none of it is to go in charity. Mankind agrees; it shall be hidden underground until he dies. Meantime Covetyse will give him further riches, through extortion of his neighbours. And now the theme changes. The time has come for Death to intervene.

> A-geyns me is no defens:
> In the gretë pestelens,
> Thanne was I wel knowe.

Mankind appeals against Death to *Mundus*, who only sends a

---

[1] dynge (digne), 'worthy'. [2] crulle (crawl).
[3] vnthende, 'weak'. [4] vnwolde (unwieldy), 'infirm'.
[5] sloppe, 'loose garment'. [6] hoppe, 'leap'.
[7] her (hair). [8] hore (hoary).

Boy, of no kin to him, to inherit his wealth. His name is 'I wot neuere who'.

> 'I wot neuere who', so wele[1] say.
> Now am I sory of my lyf:
> I haue purchasyd, many a day,
> Londys & rentys with mekyl[2] stryf;
> I haue purchasyd holt[3] & hay,[4]
> Parkis & poundys, & bouris blyfe,[5]
> Goode gardeynys with gryffys[6] gay,
> To myne chyldyr & to myn wyfe,
> In deth whanne I were dyth.[7]
> Of my purchas I may be wo;
> For, as thout,[8] it is not so,
> But a gedelynge,[9] 'I wot neuere who',
> Hath al that the Werld me be-hyth.[10]

But, as in the *Pride of Life*, the last word is not with Death. *Anima*, the soul of Mankind, arises from the bed under the castle, and appeals to the Good Angel, who refers her to Mercy. The closing scene is a Disputation of the Four Daughters of God, before *Pater Sedens in trono* himself. Mankind is called to sit at his right hand, and the last word is his *Judicium*.

> Kyng, kayser, knyt & kampyoun,[11]
> Pope, patriark, prest, & prelat in pes,
> Duke dowtyest in dede be dale & be doun,
> Lytyl & mekyl, the more & the les,
> All the statis of the werld, is at myn renoun;[12]
> To me schal thei yeve a-compt at my dygne des.[13]
> Whanne Myhel his horn blowith at my dred dom,
> The count of here conscience schal putten hem in pres,[14]
> & yelde a reknynge
> Of here space[15] whou they han spent;
> & of here trew talent,
> At my gret Jugëment
> An answere schal me brynge.

---

[1] wele (we'll).   [2] mekyl (mickle), 'much'.   [3] holt, 'copse'.
[4] hay, 'hedge'.   [5] blyfe (belive), 'quickly'.
[6] gryffys (grafts).   [7] dyth (dight), 'ordained'.   [8] thout, 'expected'.
[9] gedelynge (gadling), 'vagabond'.   [10] be-hyth, 'promised'.
[11] kampyoun (champion).   [12] at myn renoun, 'under my control'.
[13] dygne des (dais), 'dignified judgement-seat'.
[14] pres, 'difficulty'.   [15] space, 'time on earth'.

It is an effective play. The author has made a synthesis of three allegorical themes on the problems of ethics, already familiar to theologians, and given them an emphasis determined by his outlook on a society, which in the fifteenth century was showing signs of becoming an acquisitive one. He has added fresh abstractions, and here and there some psychological confusion may have been the result. Lust (*Voluptas*) is a little difficult to differentiate from Lust (*Luxuria*). Perhaps he regarded one as a *vitium* and the other as the sin to which it led. The audience would not themselves analyse too closely. To them the abstractions were presented in the likeness of living men and women. There may have been some accepted symbolism of costume to help in discriminating between good and bad characters. If the Virtues fought with roses, the Sins perhaps wielded pitchforks. A note indicates that in the last episode Mercy should be dressed in white, Righteousness in red, Truth in sad green, and Peace in black. The literary value of the play is considerable. The writer has a wide outlook upon contemporary life, and pursues his analysis of it with an unflagging wealth of illustration and range of language. A swaying action maintains dramatic interest to the end. Metrically, too, he is a competent craftsman. Normally he uses the thirteen-line stanza inherited from the miracle plays, replacing it occasionally by a similar nine-line stanza or by quatrains. He has a strong sense of rhythm, which he keeps mainly iambic, but varies with additional unstressed syllables, or emphasizes by alliteration at will. It is regrettable that we do not know his name.

The two other moralities in the Macro MS. also come from East Anglia. They had both been once the property of a monk called Hyngham. Of one there is also an imperfect copy in the Digby MS. Neither text has a title. The best guess appears to be *Wisdom*, but *Mind, Will and Understanding* has also been suggested. *Anima* is, I suppose, also a possibility. These are the names of leading characters. An opening speech by Everlasting Wisdom ends oddly with the words 'thus Wysdome begane', as if it were meant for a prologue. He is the Son, 'now Gode, now man'. *Anima*, the Soul of Man, woos him for her lover, and Wisdom analyses her two parts, sensuality and reason, whence she wears black and white. Five virgins enter, who are the five Wits, Sight, Hearing, Smell, Taste, and Touch, and with them Mind, Will, and Understanding, whom Wisdom calls the 'thre mightis' of the soul, through whom comes knowledge respectively of the Father, the Son, and the Holy Ghost. From them spring the three virtues, Faith, Hope, and Charity, and they have three enemies, the World, the Flesh, and the Devil. But of these six, the only one who appears as a character is Lucifer, who comes in 'as a prowde gallant', in the array of a devil, plotting to tempt both the soul and the flesh of man. He argues with Mind, Will, and Understanding in favour of worldly against contemplative life. They are impressed, and he triumphs. He will now stir the soul to pride, covetousness, and lechery. The episode ends rather oddly,

> Thus, by colours and false gynne,[1]
> Many a soule to hell I wyn.
> Wyde to go I may not blyne[2]
> With this fals boy; God gyff hym euell grace!
> (Her he takyt a screwd[3] boy with hym, & goth hys wey, cryenge.)

Mind, Understanding, and Will now exult over their perverted lives. They are not very well differentiated. But apparently Mind has been led by the World through Pride into sins of violence and oppression, very characteristic of the fifteenth century, Understanding by the Devil through Covetousness into sins of dishonesty, and Will by the Flesh through Lechery into sins of sensuality. These are chorographically represented. For Mind, Maintenance leads a dance of Indignation, Sturdiness, Malice, Hastiness, Wreck, and Discord. For Understanding, Perjury follows with one of six jurors, Wrong, Sleight, Doubleness, Falsehood, Ravin, and Deceit, who are said to be the Quest of Holborn, a court notorious for its venality. For Will, Lechery winds up with one of three women disguised as gallants and three of matrons, who are Recklessness, Idleness, Surfeit, Greediness, Adultery, and Fornication. Now Mind, Understanding, and Will quarrel and drive out the dancers. They discuss further exercises of their sins and jollity. Wisdom enters, reminds them of death, and shows them *Anima*, 'in the most horrybull wyse, foulere than a fende'. Six small boys, in the likeness of devils, run in and out of her mantle. The three now repent and with them *Anima*. As they leave, she sings contrition 'in the most lamentabull wyse, with drawte notys,[4]

---

[1] gynne, 'cunning'.   [2] blyne, 'cease'.   [3] screwd, 'naughty'.
[4] drawte notys, 'long-held, slow notes'.

as yt ys songyn in the passyon wyk'. Wisdom prescribes nine points of right living. *Anima*, with the five Wits, of whom we have seen nothing since the beginning of the play, and Mind, Will, and Understanding, re-enters, and the play ends in discourse with Wisdom on *Anima's* reformation and praise to Jesus. Obviously the classification of abstractions in this play is far less precise than that in the *Castle of Perseverance*, and more remote from the analysis of the theological writers. And no use is made of the traditional motives of the battle or siege, or the disputation of the Daughters of God, or the coming of Death It is probably late work. A date of about 1460 has been plausibly conjectured. It is noticeable, too, that of thirty-six characters only six take part in the dialogue. The rest are no more than decorative adjuncts. I am inclined to conjecture that the piece was intended for school performance, possibly indoors. It is clumsily written, mainly in octaves and *rime couée*, without much alliteration, but with a good deal of aureate wording.

Mankind is an example of the morality in its decadence. An Edward is King, and the play may date from about 1475. Mercy opens it, dwelling on redemption and the need for good works. She is interrupted by Mischief. A leaf seems to be missing, with an entry for Nought, New-Gyse and Now-a-Days, whom we may call the Worldlings. They dance, speak coarsely, and chaff Mercy's 'Englysch Laten'. It is noticeable that in this play aureate language is deliberately confined to the serious characters. The Worldlings do not use it. It is exaggerated, almost to the point of burlesque, and is accompanied by anapaestic unstress, but not alliteration. Mercy laments the 'vycyouse gyse' of the day. Mankind now introduces himself. His flesh and soul are contrary. Mercy gives him good advice,

> Remembur, my frende, the tyme of contynuance!
> So helpe me Gode! yt ys but a chery tyme![1]

The Worldlings return and mock. Mercy warns Mankind against them, Mischief, and Titivillus, who 'goth in-vysybull' and has a net. He is the devil whom the Wakefield Master introduced, as a Shepherd for the unjust souls, into the York *Judicium*. Here he seems to stand for Temptation. Mankind, moved by the 'mellyfluose[2] doctryne' of Mercy, betakes himself

---

[1] chery time (cherry time), 'short while'.
[2] mellyfluose, 'honey-dropping'.

to toil with his spade. The Worldlings call on 'the yemandry that ys here' to listen to a Christmas song. It is a very dirty one. They jeer at Mankind's farm labour, but he uses his spade to drive them off. Mischief consoles them. But now comes a curious 'interleccyon'.[1] A minstrel is called for, who is apparently Titivillus, and the 'worschypfull souerens' are told that the Worldlings intend to gather money, beginning with 'the goode man of this house'. The episode is rather confused. But New-Gyse and Nought are sent off to get what they can from a number of named gentry dwelling at places in Norfolk and Cambridgeshire. Meanwhile Titivillus succeeds in tempting Mankind from his honest labour, telling him that Mercy has stolen a horse, and been hanged for it in France. Mankind now makes friends of Mischief and the Worldlings and will not listen to Mercy. She laments over him. The false friends fear a reconciliation with her and try to avoid it by making him hang himself. But he survives. Mercy recovers him, shows him that the world, the flesh, and the devil had brought him to this pass, and speaks an epilogue to the 'wyrschepfyll sofereyns' of the audience. Obviously this is a very degraded type of morality, aiming at entertainment rather than edification, and well on its way towards the farce of the sixteenth century. The language, except where Mercy intervenes, is extremely coarse. The monk Hyngham ought to have been ashamed of claiming ownership. There are only about 900 lines in the play, and only seven actors are required. Dr. Pollard has plausibly suggested that it was given by a strolling company in the court-yard of an inn, and that the collection taken at one point was a real one. All spectators of medieval plays are by convention 'worshipful sovereigns'.

The Summoning of Everyman makes amends for *Mankind*. This last example of the typical fifteenth-century morality returns to the simplicity of its first surviving predecessor, *The Pride of Life*. Like that, it has a single theme, which is again the coming of Death. Exceptionally, it has reached us in printed form. There are two fragments from the press of Richard Pynson (1493–1530), and two full editions from that of John Skot (1521–37). The play is poor in metrification. It is mainly written in four-stress couplets, varied by quatrains, but the stressing is very irregular, and rhymes often fail or are imperfect.

---

[1] interleccyon, 'talk', 'consultation'.

There may well be some textual corruption. But the action is simple and solemn, and, as modern revivals have shown, dramatically effective. A long title calls the work a 'treatyse', but adds that it is 'in maner of a morall playe'. Scholars differ as to its probable relation to a Dutch *Elckerlijk*, printed about 1495, and possibly written by Peter Dorland of Diest. After a prologue by a Messenger, asking for 'audyence', God complains that the people forsake him and use the seven deadly sins. He summons Death, the 'myghty messengere', and sends him to warn Everyman that he must take upon him a pilgrimage and bring a sure reckoning. Death obeys, and finds Everyman unready, but can give him no respite. Everyman desires a companion on his journey, and appeals successively to Knowledge, to his Cousin, and to his Kindred, but in vain, nor will his Goods go with him.

> I lye here in corners, trussed and pyled so hye,
> And in chestes I am locked so fast,
> Also sacked in bagges, thou mayst se with thyn eye;
> I can not styre; in packes lowe I lye.

Moreover,

> My condycyon is, mannes soule to kylle.

Everyman turns to his Good Deeds, but she is so weak, she cannot stand on her feet. She refers him to her sister Knowledge, who offers to guide him and takes him to Confession, who dwells in the house of salvation, and gives him the scourge of penance. Everyman uses it, and now Good Deeds becomes strong enough to walk in his company. On the advice of Good Deeds and Knowledge, Everyman also calls for Strength, Discretion, Five Wits, and Beauty. They agree to accompany him, and he makes his testament, leaving half his goods in charity and the rest 'in queth[1] to be retourned there it ought to be'. Knowledge and Five Wits now advise Everyman to take the holy sacrament and unction from Priesthood. No such character appears, but the dignity of Priesthood is dwelt on at length. Everyman goes and returns, and now a cave opens into which he must creep. At this point nearly all his companions fail him. Beauty, Strength, Discretion, and Five Wits in turn refuse to enter the

---

[1] queth, 'bequest'.

cave. Knowledge will wait to see the end. Only Good Deeds remains faithful.

> All erthly thynges is but vanyté;
> Beaute, strength and dyscrecyon do man forsake,
> Folysshe frendes and kynnesmen, that fayre spake,
> All fleeth saue good dedes, and that am I.

Knowledge hears the angels sing, while Everyman is welcomed to the presence of Jesus. His reckoning is now 'crystall clere'. I am no theologian, but the strong emphasis on Good Deeds seems to me to suggest a Protestant temper rather than a Catholic one. Perhaps the long passage on priesthood and the seven sacraments was introduced as a makeweight.

In 1471 a *ludus de Bellyale* was given at Aberdeen. We have no details of it, but presumably it was based, directly or indirectly, upon the treatise *Processus Belial*, written by Jacobus Palladinus de Theramo in 1381, which incorporated the theme of the Four Daughters of God with others already, like that, used by Hugo of Saint Victor, of debates between Christ and the Devil, and between the Virgin and the Devil, and of a lawsuit, to which God, Man, and the Devil are parties. Some such performance as that at Aberdeen is suggested by a curious fragment to which has been given the name *Processus Satanae*. It is an actor's 'part' for God, the cues of which, with other indications, show that Christ, Satan, Michael, Verity, Justice, Mercy, Peace, and an Angel were also included in the cast. The actor's name, or possibly that of the scribe, was Jiggons. The manuscript, which is in a hand of about 1570–80, probably came from the Harley family of Herefordshire, but a stage representation of God would be unusual at that date, and it may be that we have only an antiquarian copy. Or can it have been given, in a remote district, by travelling Catholic actors, who were brought before the Justices of the Peace? Finally, we may note an intrusion of morality elements into the late saint play of *Mary Magdalen* already noticed. The first part of this deals with the early and reprobate life of the heroine. *Mundus*, Flesh, and Satan are there, with the Seven Sins and Sensuality, a Messenger of *Mundus*. Lechery is sent to tempt the 'beryl of beauty'. The Seven besiege her castle. Lechery enters with her Bad Angel, and takes her to a house in Jerusalem, where she is corrupted by a gallant Curiosity. The Seven then enter the house of Simon the Leper, but here a Good Angel converts Mary, and she washes the feet of Jesus, who bids her go in peace. There is of course scriptural warrant for this. The Seven and the Bad Angel go to Hell, where the *Rex Diabolus* sets Belfagour and Belzabub to beat them all on the buttocks.

Of secular medieval drama there is little to record. It cannot be supposed that the late-thirteenth-century *Interludium de Clerico et Puella* stood alone, but the omnipresent minstrels have left us little detail of their doings. The *Interludium* rests upon earlier *fabliaux* of *Dame Siriz*. We have only a fragment of forty-two octosyllabic couplets of it, making two short scenes. In the first a 'damishel' rejects the wooing of a 'clerc of scole'. She is 'mayden Malkyn'. In the second he resorts for help to Mome[1] Elvis, who professes her piety.

> Y led my lyf wit Godis love,
> Wit my roc[2] y me fede,
> Can I do non othir dede,
> Bot my pater noster and my crede.

But in the light of other versions of the story, we may suppose that he overcame her scruples. *Dux Moraud* is again an actor's 'part', preserved in a fourteenth-century hand on the margin of an assize roll. I shall always regret that I once came upon the manuscript and failed to recognize its nature. But another was more penetrating. The *Dux* boasts and bids farewell to his wife before going on a voyage. He expresses piety, but makes love to a damsel, who is, regrettably, his daughter. He fears a treacherous woman. A child is born. He bids the damsel slay it, and flees the country. At the sound of a church bell he repents. He hails his daughter, who either kills him or, in view of other versions of the theme, repents with him. The piece has been called a miracle play, but is perhaps better regarded as a *fabliau*. A London chronicle notes plays in 1444 of Eglemour and Degrebelle at St. Albans, and of 'a knight cleped[3] Florence' at Bermondsey. The titles suggest themes taken from the romances of *Sir Eglamour of Artois* and *Le Bone Florence of Rome*. A play of *King Robert of Sicily* was given at Lincoln in 1453 and another at Chester in the days of Henry VII. Here also an extant romance doubtless provided material. Two plays on Robin Hood will be best considered in relation to balladry. This is but a meagre record as compared with the rich one of religious drama.

---

[1] Mome, 'Aunt'.   [2] roc, 'distaff'.   [3] cleped, 'called'.

THE CAROL AND FIFTEENTH-CENTURY LYRIC

IF courtly poetry fell into decadence with the fifteenth-century inheritors of the Chaucerian tradition, a more popular lyric held its own, mainly by virtue of the carol. The term is of French origin, and philologists differ as to whether it owes its derivation to the Greco-Latin *chorus*, through *chorea*, a dance, or *choraules*, the flute-playing accompanist of a dance, or to *corolla*, a little crown or garland. In either case, the sense of a 'ring' is there, although the alternative 'ryng-sangis' first emerges with Gavin Douglas in the sixteenth century.

> Sum sang ryng-sangis, dansys ledis, and roundis,
> With vocis schrill, quhil all the dail[1] resoundis;
> Quhanso thai walk into thar caroling,
> For amorus lays doith the Rochys[2] ryng.

The French *carole* was a dance-song. Its beginnings have been traced by the learning of M. Gaston Paris, M. Alfred Jeanroy, and M. Jean Bédier to the twelfth century, when the courtly life of castle and manor in northern France was beginning to differentiate itself from the more homogeneous society of the eleventh century, and to develop a literature which was not as yet dominated by the *amour courtois* of Provence, with its eternal triangle of the woman, the lover, and the jealous husband. This is also the period of the poems variously called *romances*, *chansons d'histoire*, and *chansons de toile*. M. Jeanroy thinks that these themselves may have been danced. They were certainly also sung by women at their needlework. The first mention of a *carole* appears to be in the Anglo-Norman Wace's account, about 1155, of King Arthur's wedding. Here the women *carolent* and the men *behourdent*, 'jesting' while they watch the performance. So, too, in another early poem quoted by M. Jeanroy,

> Les dames main a main se tiennent,
> Et tout ainsi come elles viennent
> Se prent chacune a sa compaigne,
> Ne nus hons ne s'i accompaigne.[3]

---

[1] dail (dale).   [2] Rochys (rocks).
[3] 'Nor is any man in their company.'

But, before the end of the twelfth century, men too had begun to take part in the performance.

The earliest *caroles* do not come to us direct but only as quotations in poems of later date. They are fullest in the romance of *Guillaume de Dôle*, which often gives a whole *couplet* or stanza together with its *refrain*. *Refrain*, earlier *refrait*, comes from *refractum*, the past participle of an assumed late Latin *refrangere*, to break. Primarily, I suppose, it indicates something which qualifies what has gone before, as a wave breaks back. But in poetry it comes to have the sense of a repeated element, with some relevance to the progressive stanzas between which it recurs. In romances later than *Guillaume de Dôle*, in *pastourelles*, and in *motets*, where they are set to learned music, *refrains* often appear without the stanzas for which they are written. Their themes are mainly of love and of love from the woman's point of view, of encounters in garden or orchard, of the obstacles set by parents, of the infidelity of lovers, of absences and returns. It is so, too, in the *chansons de toile* sung by women at their needlework or weaving (*toile*). In both types of poem there is generally an element of narrative, to which emphasis is given by the recurrent lyrical refrain. The *chansons de toile* generally deal with the adventures of a Bele Erembors, a Bele Aiglentine, a Bele Yolanz, or the like. So too in the *caroles* there is constant mention of a Bele Aelis. In her story figured a Robin. According to the author of *Guillaume de Dôle*,

> Que de Robin, que d'Aaliz,
> Tant ont chanté que jusqu'as liz[1]
> Ont fetes durer les caroles.

There are many fragments of the story of Bele Aelis.

> Main se leva la bien faite Aelis;
> bel se para e plus bel se vesti:
> Si prist de l'aigue[2] en un doré[3] bacin,
> Lava sa bouche et ses oex[4] et son vis;[5]
> Si s'en entra la bele en un gardin.

But we shall never know exactly what happened to her there. An *exempla* writer of the fourteenth century had no doubt.

> Quando Aeliz de lecto surrexit, et lota fuit, et in speculo aspexit, et ornata fuit, iam cruces ad processionem tulerant et missam cantaverant, et demones eam tulerunt.

[1] liz (lit), 'bedtime'.   [2] aigue, 'water'.   [3] doré, 'gilt'.   [4] oex, 'eyes'.   [5] vis, 'face'.

A sermon, which took the *carole* for its text, gives us some notion of the dancing itself.

Cum dico Belle Aliz, scitis quod tripudium primo propter vanitatem inventum fuit: scilicet, in tripudio tria sunt necessaria, scilicet, vox sonora, nexus brachiorum, et strepitus pedum.

This may be supplemented from the twelfth-century *Gemma Animae* of Honorius of Autun, although he is professedly referring to the ancients:

Per choreas autem circuitionem voluerunt intelligi firmamenti revolutionem: per manuum complexionem, elementorum connexionem: per sonum cantantium, harmoniam planetarum resonantium: per corporis gesticulationem, signorum motionem: per plausum manuum vel pedum strepitum, tonitruorum crepitum.

A fourteenth-century *exemplum* by Jacques de Vitry adds a point:

Sicut vacca que alias precedit in collo campanam gerit, sic mulier que prima cantat et coream ducit quasi campanam dyaboli ad collum habet ligatam.

Yet another ecclesiastic writes:

Chorea enim circulus est cuius centrum est diabolus, et omnes vergunt ad sinistrum.

A miniature reproduced as frontispiece to Dr. R. L. Greene's *Early English Carols* shows a group of men and women dancers with raised arms and joined palms, moving to the accompaniment of two minstrels.

M. Jeanroy regards the themes of the *caroles* and *chansons de toile* as a courtly adaptation of those prevalent in a more popular poetry lying behind them. They are traceable also in the Bavarian *Carmina Burana*, which have much of spring and love, and the laments and exultations of women.

> Veris dulcis in tempore
> florenti stat sub arbore
> Iuliana cum sorore.
> dulcis amor!
> Qui te caret hoc tempore
> Fit vilior.

They are related to the May festivals and the burgeoning of trees.

La jus[1] desouz l'olive
— Ne vos repentez mie —
fonteine i sourt serie:[2]
puceles, carolez.
      Ne vos repentez mie
de loiaument amer.

If M. Jeanroy is right, the *caroles* may come at the end of a long
tradition, which links them ultimately with the *observatio paganorum*, against which the Church had to make head, when it
first reached western Europe. The instinct to emotional self-
expression in rhythm, as I have suggested elsewhere, finds its
outlet, not only in response to the rhythms of labour in such
folk activities as the swing of the sickle or flail or the pull of the
oar, but also in the rhythms of play, when the nervous energies,
released from the ordinary claims, are diverted into unre-
munerative channels, and under the stimulus of meat and wine
the idle feet of the chorus break into the uplifting of the dance.
This we may believe to have been notably the case at critical
seasons of the agricultural year, when our primitive ancestors
went in procession about the fields of their village, to secure
fertility to their crops and herds, gathering finally in a ring of
ecstatic dancing around some notable copse or tree, which was
regarded as the special habitation of the fertilization spirit.
And as the spirit presided over human as well as other fertility,
it was natural that women should take a leading part, and that
the impulse of the dance should be amorous. So, at least, we
may speculate. Originally the song which accompanied it may
have been no more than an inarticulate outcry. But it came to
centre round a leader, who traced a theme, while the rest, from
time to time, iterated his phrases, or later were trained to break
in at fixed intervals with a recurrent formula, which emphasized
the significance of his intention. And in this way, we may
suppose, the *carole*, with its stanza and refrain, came into existence. What is certain is that from an early date the preachers
and councils of the western Church were tilting against the
wanton songs and dances, in particular of women, which had
invaded even the sacred edifices themselves or their precincts.
This was largely due to the bad tactics of the Church itself.
An example may be found in the letters of Gregory the Great

[1] jus (jeu), 'sport'.
[2] sourt serie, 'flows serenely'.

at the time of the mission of St. Augustine to England early in the seventh century. He had bidden the converted Ethelbert of Kent to suppress the worship of idols and throw down their fanes. And then, by an afterthought, he wrote, 'Do not pull down the fanes; let them become temples of the true God. So the people will have no need to change their places of concourse, and where of old they were wont to sacrifice cattle to demons, thither let them continue to resort on the day of the Saint to whom the church is dedicated.' But the fanes were the direct descendants of the sacred trees, and when the people came to the churches they brought their dances with them, ringing the sacred edifices with amorous ditties, as they had formerly ringed the trees. As late as the end of the twelfth century Giraldus Cambrensis, in his *Gemma Ecclesiastica*, tells the story of a priest in Worcestershire who caused scandal by intoning at Mass, not *Dominus Vobiseum*, but *Swete lemman, thin are*,[1] a *particula cantilenae*, called *refectoria seu refractoria*, which he had heard all the night before *in choreis circiter ecclesiam ductis*. A widespread story of the Dancers of Kölbigk, which may have originated in Saxony or Lorraine, seems to have been disseminated by travelling mendicants afflicted by St. Vitus's dance, who claimed to be themselves survivors of the catastrophe. According to the latest and fullest version of this, a group of Christmas revellers, led by one Bovo, *tam etate prior quam stulticia*, came to a church and sent two girls, Merswinda and Wibecina, to bring out the priest's daughter Ava to join them. They linked hands and made ready their *corolla* in the porch. One Gerlevus started their *fatale carmen*:

> Equitabat Bovo per silvam frondosam,
> Ducebat secum Merswyndam formosam,
> Quid stamus, cur non imus?[2]

The priest cursed them, and their hands remained locked together for a twelvemonth. He sent his son Azon to bring Ava away, but her arm came off when he pulled it. At the end of the year she died, but the arm would not stay in the grave. The fragment of the song looks much like the opening of a *carole* or *chanson de toile*, with its *formosa* echoing the constant *'Bele'*. M. Gaston Paris cites as a parallel a *carole* in *Guillaume*

---

[1] 'Sweet mistress, thy grace'.
[2] The line is borrowed from Terence.

de Dôle, which begins:

> Rainauz o s'amie chevauchent par un pré;
> Tote nuit chevauchent jusqu'au jor cler.
> Je n'avrai jamais joie de vos amer!

The shortness of the Bovo stanza suggests a *carole* of an early type. M. Jeanroy traces a structural development in the fourteenth century from a triplet stanza in assonance or monorhyme, with a refrain in one or two lines not linked to it by rhyme. This was elaborated by the addition to the stanza of a tail line, rhymed not to its triplet but to one or both lines of the refrain. Later modifications produced in succession the *ballete*, the *rondet*, and the *virelai*, which is the form adopted in Chaucer's *Merciles Beaute*, although it is called a roundel. But these forms belong to courtly poetry, which is remote from the English popular carol.

The misfortune which befell a Worcestershire priest is not the only indication of the prevalence of amorous dance-song in this island. William Fitzstephen, towards the end of the twelfth century, records that in London 'puellarum Cytherea ducit choros usque imminentem lunam, et pede libero pulsatur tellus'. But tags from Horace do not tell us very much. Walter de Chanteloup in 1240 forbade *ludos de Rege et Regina* in the diocese of Worcester, and Robert Grosseteste about the same time 'ludos quos vocant inductionem Maii sive Autumni', which were the same thing. A decree by the University of Oxford, a little later, goes into more detail:

> Ne quis choreas cum larvis seu strepitu aliquo in ecclesiis vel plateis ducat, vel sertatus, vel coronatus corona ex foliis arborum, vel florum vel aliunde composita, alicubi incedat prohibemus.

A record of an event of 1282 preserved in the Scottish *Chronicle of Lanercost* explicitly recognizes the relation of such proceedings to the pagan ritual:

> Sacredos parochialis, nomine Johannes, Priapi prophana parans, congregatis ex villa puellulis, cogebat eas, choreis factis, Libero patri circuire.

Priapus-Liber may reasonably be identified with the Teutonic Freyr. But although these prohibitions are testimony to the prevalence of the popular dance, they are not very illuminating as to how it was conducted. We do not even get a vernacular name for it. Probably it was simply called a 'ring'. Again a sermon, quoted by Dr. Greene, helps us:

Mi leue frend, wilde wimmen & golme[1] i mi contreie, wan he gon o the ring, among manie othere songis, that litil ben wort that tei singin, so sein thei thus:

> 'Atte wrastlinge my lemman[2] i ches,'[3]
> And atte ston-kasting i him for-les.[4]

Later, in the *Stanzaic Life of Christ*, about 1327, we get:

> Thow in tho ryng of carolyng
> Spredis thin armes furth from the.

And still later, about 1450, Sir Richard Holland writes in his *Howlat*:

> Fair ladyis in ryngis,
> Knyghtis in caralyngis,
> Boith dansis and syngis.

Incidentally, reflecting, as one should do, on the sermon, we may ask what stone-casting had to do with a dance-song. The conjunction recurs in a decree, as late as 1384, by William of Wykeham, Bishop of Winchester, which condemns the pollution of graveyards, alike by dissolute dances and by stone-castings. Were these, in the precinct of a fane, merely an athletic exercise, or did they owe their origin to the pagan ritual, in the heaping up of stones upon a funeral cairn?

This, however, is a digression. The chroniclers do not tell us much about the nature of the folk-festivals. They do, however, sometimes record actual or legendary incidents of an historical character which became the subjects of poetry and were occasionally taken up into popular dance and song, as a variation from the amorous themes. William of Malmesbury, about 1125, apparently using a somewhat earlier account by Fabricius, Bishop of Abingdon, says that Ealdhelm, in the seventh century, wrote songs which were pleasing to King Alfred, and was moreover wont to stand on a bridge and waylay the folk to piety, 'quasi artem cantitandi professum, sensim inter ludicra verbis scripturarum insertis'. One such *carmen*

---

[1] golme (?), 'wanton' (?).
[3] ches (chose).
[2] lemman, 'lover'.
[4] for-les, 'lost'.

triviale had endured to William of Malmesbury's own time. From the *Liber Eliensis*, also of the twelfth century, we get the opening words of a song said to have been composed by King Canute (1013–35), 'quae usque hodie in choris publice cantantur et in proverbiis memorantur':

> Merie sungen the muneches binnen Ely,
> Tha Cnut ching reu ther by;
> Roweth, cnites, noer the land,
> And here we thes muneches sæng.[1]

No refrain is apparent, and conceivably this was originally a boating song, perhaps later adapted for the purposes of *chori*. It may have been narrative, rather than lyric, and is further discussed in Chapter III. From William of Malmesbury we get a story of the salvation of Canute's daughter Gunhild from a false accusation of adultery, which again was *nostro adhuc saeculo in triviis cantitata*. The exploits of Hereward the Saxon about 1070 are said in his *Gesta* to have been recorded in fables by his priest Leofric, but in the *Chronicle of Croyland*, a forgery of the late fourteenth century, it is added that women and maidens sang of them in their dance. It is William of Malmesbury again who tells us that stories of Athelstan (895–940) were preserved in his day 'magis cantilenis per successiones temporum detritis, quam libris ad instructiones posterum'. Layamon, in his *Brut* of about 1189, tells us that after an Arthurian battle,

> Tha weoren in thissen lande blisfulle songes.

In the thirteenth century Matthew Paris, the chronographer of St. Albans, cites two lines ascribed to Flemish mercenaries of the Earl of Leicester in revolt against Henry II:

> Hoppe,[2] hoppe, Wileken, hoppe, Wileken,
> Engelond is min ant tin.[3]

At the beginning of the fourteenth, Edward I, invading Scotland, was met by seven women, 'obviantibus regem per viam, et cantantibus coram eundem', as they had been wont to do in the reign of King Alexander. A Scottish song, made by

---

[1] Merrily sang the monks by Ely, when King Canute rowed thereby. 'Row ye, my knights, near to the land, and hear we the singing of these monks.'
[2] Hoppe, 'dance'.   [3] ant tin, 'and thine'.

maidens after the battle of Bannockburn in 1314, is recorded
in the chronicle known as the *Brut*:

> Maydenes of Engelande, sare may ye morne,
> For tynt[1] ye have youre lemmans[2] at Bannokesborn,
>   With hevalogh.[3]
> What wende[4] the Kyng of Engeland
> To have ygete[5] Scotlande
>   With rombylogh.[6]

This is reproduced, with some variation of wording, by Robert
Fabyan, in his *New Chronicles* of 1516, who adds that it was
'after many dayes sungyn, in daunces, in carolis of ye maydens
& mynstrellys of Scotlande'. But again it must have been
originally a rowing song, in view of other such songs which use
'rombelow' as a refrain. John Barbour in 1375, describing
another victory by the Scots, half a century before, says:

> I will nocht reherse all the maner;
> For quha sa likis, thai may heir
> Young women, quhen thai will play,
> Syng it emang thame ilke day.

There is, of course, a strong element of minstrelsy in all this.
In fact, the precise story of Gunhild's trial, transferred to her
mother Emma, was sung by a *joculator* called Herebertus in the
hall of the Prior of St. Swithin's monastery at Winchester,
during 1338. But dancing women may well have varied their
traditional themes of song from time to time by others taken
from heroic lays to which they had listened.

'Carol' first appears as an English term just before the end
of the thirteenth century, in the *Cursor Mundi*. Here a speaker
says,

> Caroles, iolités, and plaies,
> ic haue be-haldyn and ledde in ways.

And in another passage we learn

> To ierusalem that heued[7] bare thai,
> Ther caroled wiues bi the way;
> Of thair carol suche was the sange,
> Atte[8] thai for ioy had ham amange.

---

[1] tynt, 'lost'.　　　　　　　　　　　　　　[2] lemmans, 'lovers'.
[3] hevalogh, a boating refrain='Heave-ho!'　　　　[4] wende, 'thought'.
[5] ygete, 'got'.　　[6] rombylogh (rumbelow), another boating refrain.
[7] heued (head).　　　　　　　　　　　　[8] Atte, 'that'.

This suggests a processional performance, rather than a ring. We can hardly think of pilgrims stopping to dance round the head of a martyr on the dusty roads of Palestine. A little later Robert Mannyng of Brunne, translating William of Wadington's *Manuel de Péchiez* in his *Handlyng Synne*, writes:

> Karolles, wrastlynges or somour games,
> Whoso euer haunteth any swyche shames
> Yn cherche, oder yn cherche3erd,
> Of sacrylage he may be a ferd.

In a version of the story of the Dancers of Kölbigk, taken from a source other than that of William of Wadington, he has 'Karolle' both as noun and verb, 'Karollyng', 'Karolland'. He criticizes women who borrow clothes 'yn carol to go'. And for him Stonehenge is 'the Karolle of the stones'. So too in Michael of Northgate's *Ayenbite of Inwyt* of 1340 we get:

> Oure blisse is y-went in-to wop,[1] our karoles into zor3e.[2]

The word is now fully naturalized. It is used of popular dancing in the romance of *Arthour and Merlin*:

> Miri time it is in May,
> Than wexeth along the day,
> Flowers schewen her borioun,[3]
> Miri it is in feld & toun,
> Foules miri in wode gredeth,[4]
> Damisels carols ledeth.

In *King Alisaunder* a similar reference is, as in France, to courtly performance. So it is too, later, in *Sir Gawayn and the Grene Kny3t*. When King Arthur kept his Christmas court with the Round Table at Camelot,

> Justed ful jolilé thise gentyle kni3tes,
> Sythen kayred to the court caroles to make.

After a challenge by the Green Knight to Gawain, Arthur bids the queen not to be perturbed:

> Wel bycommes such craft vpon Cristmasse,
> Laykyng[5] of enterlude3, to la3e[6] and to syng,
> Among thise kynde caroles of kny3te3 and ladye3.

---

[1] wop, 'weeping'. [2] zor3e, 'sorrow'. [3] borioun (burgeon), 'bud'.
[4] gredeth, 'call'. [5] Laykyng, 'Playing'. [6] la3e (laugh).

On St. John's day the company

> Daunsed full dreȝly[1] wyth dere caroleȝ.

A year later Gawain rode to take up the challenge, and was entertained in a castle.

> Much glam[2] and gle glent[3] vp therinne
> Aboute the fyre vpon flet,[4] and on fele[5] wyse
> At the soper and after, mony athel[6] songeȝ,
> As coundutes of Krystmasse and caroleȝ newe.

A 'coundute', however, is not a dance-song but a part-song, properly a processional chant (*conductus*). Gawain has himself assoiled before his coming adventure,

> And sythen he mace[7] hym as mery among the fre ladyes,
> With comlych caroles and alle kynnes[8] ioye,
> As neuer he did bot that daye, to the derk nyȝt,
>   with blys.

The carol is also still an element in the welcome of sovereigns. It met Edward II on his return to London from France in 1308, and took place in church and street at the birth of his son Edward in 1312. Here the word can stand for little more than general rejoicing. Clerical criticism of the dancing continues. I have already quoted the *Stanzaic Life of Christ* of 1327, with its reference to the spreading of arms in the ring. Another gesture is noted about 1400 in a religious poem on Jesus and the Worldling.

> A-cros thou berest thyn armes,
> Whan thou dauncest narewe;[9]
> To me hastou non awe
> But to worldés glorie.

John de Bromyard, in his *Summa Predicantium* of about 1360, has much to say. His terms are *tripudium, chorea, cantilena*. Women with garlands are the devil's packhorses for sale. Christmas has become only a feast of words, with dancing and ditties. He echoes the French preachers.

> Dyabolus se habet ad modum porcarii qui, uolens omnes porcas similiter congregare, facit unam clamare, sic dyabolus facit unam cantilenam incipere quam ipsemet dictauit.

---

[1] dreȝly, 'lengthily'.   [2] glam, 'noise of merriment'.   [3] glent, 'started'.
[4] flet, 'floor'.   [5] fele, 'many'.   [6] athel, 'noble'.   [7] mace (makes).
[8] kynnes (kinds).   [9] narewe (narrow), 'closely'.

Other moralists follow him, well into the fifteenth century. John Mirk, in his *Festial*, puts his point in the form of a *planctus*, or lament, from the Cross.

> Thou hast thyn armes sprad on brode ledyng Carallys, and I for thy love haue myn armes sprad on the tre and tachut[1] wyth gret nayles.

For the author of *Jacob's Well* it is 'a sin of dede to usyn karollys and dauncys'.

At an early stage in the development of the French *carole*, it became customary to link stanza and refrain by a common rhyme, the recurrence of which at the end of each stanza gave a signal to the throng to break into the refrain. With this clue we may search the surviving examples of late-thirteenth-century or fourteenth-century English lyrics for traces of dance-song. There are not many of them, and they are not all free from ambiguity. The well-known 'Sumer is icumen in' (B.M. Harleian MS. 978, *c.* 1240) has a refrain, and uses a seasonal theme, but in the form in which it has come to us it is a part-song for learned musicians. Another early piece (Lincoln's Inn, Hale MS. 135, *c.* 1300) has both spring and love in its refrain.

> Nou sprinkes the sprai,
>   All for loue icche am so seeke
> That slepen i ne mai.

It is personal rather than communal utterance, both in the refrain and in the stanzas, which begin like a *chanson d'aventure*,

> Als i me rode this endre dai
> O mi pleyinge.

But the rhyme-link is there, and the subordination of the throng to the leader is a natural tendency in the development of lyric. Two curious fragments come from the flyleaf of a manuscript (Bodleian Rawlinson MS. D. 913, *c.* 1300), which also contains some minstrelsy. One may be the beginning of a dance-song, with its refrain, as is often the case, prefixed.

> Icham of Irlaunde,
> Ant of the holy londe
> Of Irlande.

---

[1] tachut (attached), 'fastened'.

> Gode sire, pray ich the,
> For of saynte charité,
> Come and daunce wyt me
> In Irlaunde.

The other, although it has an element of refrain, in such phrases as

> Maiden in the mor lay,
> In the mor lay,

reads more like a dialogue between two singers, of whom one echoes the other in similar words. It may be added that minstrels, such as the author of *A Song of Lewes* and later Laurence Minot, sometimes borrowed the method of popular song, and ended their stanzas with a refrain or what might be called a quasi-refrain, with some variation of wording. A *Song of Lewes* is in a notable manuscript (B.M. Harleian MS. 2253, *c.* 1320), possibly written at Leominster Priory, which also contains a number of lyrics, both secular and religious, mostly written towards the end of the thirteenth century. Of these three, all with the personal note of the individual singer in them, have refrains. That in *Alysoun* ends with the woman's name, to which the stanzas are rhyme-linked. The second has no rhyme-link, but the refrain is written as a heading, and its repetition indicated.

> Blow, northerne wynd,
> Sent thou me my suetyng.[1]
> Blow, northerne wynd,
> Blou! blou! blou!

The third is the most interesting. It begins:

> Lutel wot hit anymon,
> How derne[2] loue may stonde.
> Bote hit were a fre wymmon,
> That muche of loue had fonde.

The last stanza is charming:

> Ffayrest fode[3] vpo loft,[4]
> My gode luef, y the greete,
> Ase fele sythe[5] & oft,
> As dewes dropes beth weete,
> Ase sterres beth in welkne,[6] ant grases sour ant suete;
> Whose loueth vntrewe, his herte is seldé[7] seete.[8]

[1] suetyng (sweeting).   [2] derne, 'deep'.   [3] fode, 'creature'.
[4] vpo loft, 'exalted'.   [5] fele sythe, 'many times'.   [6] welkne (welkin), 'sky'.
[7] seldé (seldom).   [8] seete (set), 'fixed'.

The refrain, again unlinked, is

> Euer & oo for my leof icham in grete thohte,
> Y thenche on hire that y ne seo nout ofte.

And on the same page of the manuscript is what can only be called a religious parody of this:

> Lutel wot hit anymon,
>   Hou loue hym haueth ybounde,
> That for vs o the rode ron,[1]
>   Ant bohte vs with is wounde.
> The loue of hym vs haueth ymaked sounde,
> Ant ycast the grimly gost to grounde.

And for refrain,

> Euer & oo, nyht & day, he haueth vs in is thohte,
> He nul nout leose that he so deore bohte.

These are anonymous pieces and can only be approximately dated. The Leominster collection may represent a gathering over many years, and possibly contains some thirteenth-century work. Identifiable authors begin with William Herebert, who died in 1333, and speaks of his own poems (Phillipps MS. 8336) as loose translations. Three of them have refrains, in two cases rhyme-linked. One, in *rime couée*, on the vanity of life and its end in death, is more or less translated from the Anglo-Norman of Nicholas Bozon. Another is from a Latin Palm Sunday hymn, and the third, a lament, from the service for Good Friday. More singable than any of these is another *rime couée* lyric, datable about 1372, by Johan de Grimestone (National Library of Scotland MS. Advocates 18.7.21). It is an address to Christ on His sufferings and those of His mother, but has a refrain, not rhyme-linked, which may be an adaptation from popular song.

> Luueli ter of loueli eyghe,
>   Qui dostu me so wo?
> Sorful ter of sorful eyghe,
>   Thu brekst myn hert a-to.

One other theme in this early poetry is of special interest: It is that of the lullaby. The first example (B.M. Harleian MS.

---

[1] ron (ran), 'went'.

913), not later than the first half of the fourteenth century, begins:

> Lollai, lollai, litil child, whi whepistou so sore?
> Nedis mostou wepe, hit was iȝarkid[1] the ȝore[2]
> Euer to lib[3] in sorow, and sich and mourne euere,
> As thin eldren did er this, whil hi aliues were.
> Lollai, lollai, litil child, child, lolai, lullow.
> In-to vncuth[4] world icommen so ertow [5]

There is only a quasi-refrain, opening after each stanza with 'Lollai, lollai, litil child'. It must be remembered that the dance was not the only primitive activity, the rhythm of which evoked that of song. The rocking of the cradle was another. This is not in the full sense a religious poem. In foretelling a life of sorrow for the human child, it refers, indeed, to the sin of Adam, as well as to the pre-Christian wheel of Fortune, but there is nothing of divine redemption. But it was clearly the model for another lullaby, which is one of three in Grimestone's manuscript. This is sung, not by a human mother, but by the Virgin to the Christ-child, and anticipates the Crucifixion. It has the same stanza-form as the earlier poem, and a similar quasi-refrain, also beginning 'Lullay, lullay, litel child'. Grimestone's second lullaby, again by the Virgin, is in simpler form, a *rime couée* triplet, with a rhyme-link to the true refrain, borrowed from the Harleian MS. poem, 'Lullay, lullay, litel child, qui wepest thu so sore?' His third, of thirty-seven quatrains, is too long for singing, but again has a refrain, not rhyme-linked, 'Lullay, lullay, la lullay, mi dere moder, lullay'. Finally a sermon of about 1350 (Bodley MS. 26) incorporates a poem, which, although again religious, has in its refrain a curious echo of the early connexion between song and dance:

> Honnd by honnd we schulle ous take,
> And joye and blisse schulle we make,
> For the deuel of elle man haght forsake,
> And Godes Sone ys maked oure make.[6]

(1)

> A child is boren amonges man,
> And in that child was no wam;[7]
> That child ys God, that child is man,
> And in that child oure lif bygan.

---

[1] iȝarkid, 'appointed'.   [2] ȝore, 'of old'.   [3] lib (live).   [4] vncuth, 'unknown'.
[5] ertow, 'art thou'.   [6] make, 'mate'.   [7] wam, 'blemish'.

(2)

Senful man, be blithe and glad,
For your mariage thy peys ys grad,[1]
    Wan Crist was boren;
Com to Crist; they peis ys grad;
For the was hys blod ysched,
    That were forloren.

(3)

Senful man, be blithe and bold,
For euene[2] ys bothe boght and sold,
    Euereche fote.
Com to Crist; thy peys ys told,
For the he yahf[3] a hondrefold
    Hys lif to bote.

The repetition of the refrain is indicated after each stanza by quoting its opening words.

These refrain poems are of a transitional character. They indicate a movement towards the sanctification of popular song, which began in the fourteenth century, but did not reach its full fruition until the fifteenth. I feel no doubt that Dr. Greene is right in ascribing its main impulse to the literary activity of the English Franciscans. St. Francis himself had brought a note of humanity and even gaiety into the Italian religious revival of the thirteenth century. His Christmas crib at Greccio may have been merely a revival of an older custom. But he certainly bade his companions become *ioculatores Dei* and mingle *laude* with their preaching, and his tradition is carried on in the *laude* of Jacopone da Todi and the companies of *laudesi* who followed. The Franciscans came to these islands in 1224. Already about 1275 Thomas of Hales, who was of the Order, had written a pious song (Jesus Coll. Oxf. MS. 29) for a nun, who asked of him a 'luue ron'.[4] And of the fourteenth-century refrain pieces just described several are of similar origin. Nicholas Bozon, William Herebert, Johan de Grimestone were all Franciscans. The earliest lullaby comes from a Franciscan manuscript, the 'Honnd by honnd' poem from a Franciscan sermon. Even more illuminating is the 'Red Book of Ossory', which Dr. Greene cites from a manuscript in the

---

[1] grad, 'proclaimed'.   [2] euene (heaven).
[3] yahf (gave).   [4] luue ron, 'love-song'.

diocesan palace of Kilkenny, and is now engaged in editing.
It contains a number of lyrics composed between 1317 and
1360 by Bishop Richard de Ledrede, a Franciscan, for the use
of the minor clergy of his cathedral, 'ne guttura eorum et ora
deo sanctificata polluantur cantilenis teatralibus turpibus et
secularibus'. The term *teatralibus* suggests a condemnation
rather of minstrelsy than of popular song. The lyrics themselves are in Latin, often with a refrain. But to each is prefixed
a scrap of English or Anglo-Norman verse, to the tune of which
it was to be sung. Dr. Greene gives some examples:

> Alas hou shold Y synge? Yloren[1] is my playnge.
> Hou shold Y with that olde man
> To leuen, and let[2] my leman,[3]
>   Swettist of all thinge?

> Do, do[4] nightyngale synges ful myrie
> Shal Y neure for thyn loue lengré[5] karie.[6]

> Haue God day my lemon[3] &c.

> Heu alas par amor
> Qy moy myst en taunt dolour.

I think the good bishop must have been concerned with the
reformation of an abuse prevalent in ecclesiastical foundations
under the name of the Feast of Fools. It was common, towards
the end of the tenth century, to regard the *triduum*, which
followed the Nativity, as a period in which special honour
should be done to the minor clergy of these establishments.
On St. Stephen's Day the deacons, on St. John's Day the
priests, on Innocents' Day the choir-boys, were allowed, as
we learn from the *Rationale Divinorum Officiorum* of Joannes
Belethus, to undertake the whole conduct of the services, with
the exception of the Mass itself. And to these *tripudia*, as
Belethus calls them, had been added, by the end of the twelfth
century, apparently with less explicit authority, a fourth,
variously dated at the Circumcision, the Epiphany, or its
octave, in which the same privilege was given to the sub-
deacons. Their feast was conducted with less sobriety than the
others, and often degenerated into a scandalous riot, under
the name of the *festum fatuorum* or the *asinarium festum*. The
practice was particularly widespread in France, and attempts

---

[1] Yloren, 'lost'.   [2] let, 'leave'   [3] leman, lemon, 'lover'.
[4] Do, do, perhaps musical notes.   [5] lengré (longer).   [6] karie (care), 'mourn'.

to reform it led to a condemnation by Innocent III in 1207, which was incorporated in the *Decretals* of Gregory IX in 1234. It proved, however, ineradicable, and some of its extravagances affected the older feasts of the *triduum* itself. In England it may have been less widespread. But it called for prohibitions at Lincoln by Bishop Grosseteste in 1236 and Archbishop Courtney in 1390, and apparently still survived as *ly ffolcfeste* in 1437. Its extirpation is required also by the statutes of Beverley Minster in 1391. Richard de Ledrede may very well have been attempting to keep it within the bounds of decency. That Christmas should become a purely religious anniversary was doubtless impossible. It was Yule, as well as Christmas, and to it had shifted many folk-customs which, like those of May Day or Midsummer, belonged to the pagan observance. Yule had once come in mid-November, when the first fall of snow announced that a new agricultural year was beginning, and when the Roman new year in midwinter was adopted, the pagan practices were transferred to it. Everything had to be done at Yule to make the coming twelve months fortunate. There was banqueting, and tables were spread for the spirits of ancestors, who were abroad, to take their share. Wassail was drunk in ceremony, and cider poured on the roots of apple-trees. New fire was brought into the house with the Yule log, and its ashes mingled with the seed corn. Neither fire nor iron might be taken out. The first foot to enter the house must be of good omen, that of a man, preferably a dark man of the primitive strain, not that of a woman, and above all not that of a celibate priest, which would ruin the hope of children. There are traces of sacrifice, in the ceremonial eating of the boar, once sacred to Freyr, the bleeding of animals, the hunting of the wren or squirrel; even of human sacrifice, in the whipping of boys, the clothing of men in beast-skins as hobby-horses or Christmas bulls, the election of mock kings, once destined to be slain when their brief reign was over. The fertilization spirit was honoured, not as in spring with may blossoms, but with decorations of holly, ivy, and mistletoe, which bear berries conspicuous in winter, and by the interchange of gifts, *strenae* or *étrennes*, originally twigs plucked from a sacred grove. No doubt, in the course of centuries the primitive significance of such customs had been forgotten. They were at most looked upon as bringing good luck. But throughout the Middle Ages

Christmas continued to be a great secular feast as well as a religious one. There was banqueting already in Anglo-Saxon times. William the Conqueror in 1066 chose Christmas Day for his coronation, as Charlemagne had done for his in 800. From 1085 the Saxon Chronicle regularly records where the anniversary was kept. Traces of sacrifice are to be found in the masked figures, whose *ludi* or 'mummings' ultimately took a quasi-dramatic form, and in the mock king, who led the revels. He is the *Rex Fabae* in 1334. Later he became the Lord of Misrule. The example of the court was followed in noble houses and in the colleges of Oxford. It is significant of the relation between Christmas and Yule that the secular feast began, not at the Nativity, but at All Saints' Day on 1 November. This we learn from the *Liber Niger* of Henry VI, about 1478, in an account of the Chapel Royal.

The King hathe a song before hym in his hall or chambre uppon All-hallowen day at the latter graces, by some of these clerkes and children of chappel in remembrance of Christmasse; and soe of men and children in Christmasse thorowoute. But after the songe on All-hallowen day is done, the Steward and Thesaurere of houshold shall be warned where it liketh the King to kepe his Christmasse.

There is confirmation by John Stow in his *Survey of London* (1598). He is writing apparently of the fifteenth century.

In the feaste of Christmas, there was in the kinges house, wheresoeuer he was lodged, a Lord of Misrule, or Maister of merry disports, and the like had yee in the house of euery noble man, of honor, or good worshippe, were he spiritual or temporall. Amongst the which the Mayor of London, and eyther of the shiriffes had their seuerall Lordes of Misrule, euer contending without quarrell or offence, who should make the rarest pastimes to delight the Beholders. These Lordes beginning their rule on Alhollon Eue, continued the same till the morrow after the Feast of the Purification, commonlie called Candlemas day.

The All Saints' Day song of 1478 is not called a carol. It might have been an anthem. Tudor household accounts show rewards for singing *Audivi* on All Saints' Day and *Gloria in Excelsis* at Christmas, and to the Boy Bishop at the time of his election on St. Nicholas Day, 6 December, to lead the services on that of the Holy Innocents. On the other hand, we have the record of a banquet on Twelfth Night, 1487, when

At the Table in the Medell of the Hall sat the Deane and those of the Kings Chapell, which incontynently after the Kings furst course sange a Carall.

William Newark was rewarded for the making of a song at the Christmas of 1492, and either John or William Cornish by Elizabeth, the queen of Henry VII, in 1502 'for setting of a Carrall vpon Cristmas Day'.

There was certainly much dance and song at the secular feast. Gawain's courtly carols on Christmas and St. John's Days may have been harmless enough. But the revelry often took a turn inconsistent with the solemnity of the ecclesiastical commemoration. The Wyclifite *Ave Maria* allows of dancing in measure and 'honeste songis of cristis incarnation, passion, resurexion & ascencion, & of the ioies of oure ladi', but adds that 'nowe he that kan best pleie a pagyn of the deuyl, syngynge songis of lecherie, of batailis and of lesyngis, is holdyn most merie mon & schal haue most thank of pore & riche; & this is clepid worschipe of the grete solemnyte of cristismasse'. Thomas Gascoigne, a little later, in his *Loci e Libro Veritatum*, says that he knew a man who once heard an indecent song at Christmas and not long after died of a melancholy. The Christmas song in the play of *Mankind* is enough to give anyone a melancholy. It is not, however, courtly or in form a carol, but a cumulative poem with a final chorus. In the fifteenth century the author of a poem on the favourite theme of *Timor Mortis* writes, not indeed with specific reference to Christmas,

> Whe schold neuer lust, hop, ne dawnce,
> Nother syng no song of this new ordenance,
> As 'Hart myne, why nelt thou glad and lusti be?'
> Or ellys, 'ma bel amour, ma ioy en esperance',
> But sey 'Timor mortis conturbat me'.

I suspect that 'lust' in the first line should be 'joust'. This injunction is austere. But a further development of the experiments of the Franciscans, in an attempt to bring the revelry of Christmas into closer accord with the temper of the ecclesiastical feast, is the best explanation of the emergence in the fifteenth century of a large number of refrain poems, hilarious but at the same time religious in character, which seem to have inherited alike the name and the metrical form of the danced carols. They were not, however, danced, but merely sung in company. A whole household could take part in them. John Awdelay,

the earliest writer of them whom we can identify, heads a group
which he had composed with the couplet,

> J pray ȝow, syrus, boothe moore and las,[1]
> Syng these caroles in Cristemas.

The distinction between the leader who sang the successive
stanzas, and the throng, who joined after each in the refrain,
endured. But by the end of the century all were expected to be
able to act as leaders.

> Therefore euery mon that ys here
> Synge a caroll on hys manere;
> Yf he con[2] non we schall hym lere,[3]
> So that we be meré allway.

Musical notation in manuscripts shows that, with the development of choirs, carols sometimes came to be rendered as part-
songs. The division between stanza and refrain, however,
remained fundamental. Galfridus Grammaticus, in his *Promp-
torium Parvulorum* (1440), glosses *caral* as 'songe' and '*palino-
dium*', doubtless in its primitive sense of 'repetition', not the
later one of 'recantation'. The *Catholicon Anglicum* (1483) gives
'corea, chorus' for 'caralle', but also for 'dawnce'. Dr. Greene
would substitute for 'refrain' the term 'burden', keeping
'refrain' for the secondary repeated line which forms part of the
stanza. But chronology is against him here. The first example
of 'burden', in this sense, is of 1589, and of 'foot', which is
another alternative, also in the sixteenth century. 'Refrain',
on the other hand, is Chaucerian:

> But evere mo 'Allas!' was his refreyn.

The French term *couplet* for the varied part of a *carole* did not
get into English use in this sense. Probably its earliest equivalent is simply 'verse', but this is a little ambiguous, as 'verse'
primarily signifies a single line. The 'baston' of *Cursor Mundi*
did not establish itself. 'Stave', of which 'staff' is a late variant,
emerges in the fifteenth century, and 'stanza' in the sixteenth.
In *Love's Labour's Lost* Holofernes says 'Let me hear a staff, a
stanza, a verse; lege, domine'.

Carols, no doubt, came in course of time to be sung on other

---

[1] las (less). [2] con, 'know'. [3] lere (learn), 'teach'.

occasions than Christmas, but, whatever their subject-matter, collections of them are always described by sixteenth-century printers as 'Christmas carols'. Perhaps the domination of the winter festival is a little obscured in the arrangement of Dr. Greene's magnificent collection. Here the carols, 474 in number, are classified under forty-two heads. A few of them, datable before the fifteenth century, I have regarded as transitional in character. On the other hand, a few can be added from sources which Dr. Greene has not used. Of the 474, he groups only ninety-two as 'Carols of the Nativity'. Christmas, however, even as an ecclesiastical season, must be taken in a wide sense. It began on Christmas Eve, with the singing of 'Farewell, Advent', and it ended on Candlemas, with that of 'Good day, Sir Christmas'. Dr. Greene's groups, for the whole period, include 137 carols. I think we ought to add the fourteen 'Lullaby' carols, some of which definitely represent the Virgin as singing on 'Yolis day', and the twenty-four 'Annunciation' carols, which often refer to the divine birth as an accomplished event. Annunciation and Nativity are indeed hardly separable as subjects for song or narrative. On similar grounds I should add a few at least of Dr. Greene's fifty-seven 'Carols to the Virgin'. If I am right, Christmas will by itself account for well over a third of the total collection. I agree, however, with Dr. Greene that the occurrence of the word 'Nowell' in a carol-refrain is not by itself an indication of Christmas usage. It is no doubt derived from *natalis*, but had long come to signify nothing more than a general expression of rejoicing. The author of one carol has 'nova, nova' in his refrain, and perhaps thought that to be the significance of 'Nowell'. There are, of course, a large number of religious carols, devotional or didactic, which would be appropriate either at Christmas or on other occasions. Three of them, in one manuscript, are indicated as to be sung *in die nativitatis*, and one, on Purgatory, *in fine nativitatis*. For the rest, the note is merely *ad placitum*. There is no evidence of any regular practice of carol-singing at the feast of the Annunciation on 25 March, or at Easter or Whitsuntide. Carols in honour of saints, outside those of Christmas in the ecclesiastical sense, are rare. One on the Trinity and All Saints may have been sung on All Saints' Day, when the festal season at court began. It makes mention of 'Edward Kyng', presumably the Confessor, whose body lay in St. Stephen's Chapel, hard by

Westminster Hall, where the royal banquets were held. Another, on St. Edmund, refers to him as

> that worthi kyng,
> That sufferid ded this same day,

and may have been sung at Bury St. Edmunds on 20 November, the anniversary of his martyrdom. Of two on St. Nicholas, one might come at Christmas, in connexion with the ceremony of the Boy Bishop on Innocents' Day, which long survived the Feast of Fools. The other, addressed to 'maydenis', would be less appropriate then. One on St. Catherine might be for her day, 25 November. John Awdelay has three carols of saints, but of these two were apparently not meant for singing at all. Both are unusually long, and each ends with practically the same stanza,

> I pray youe, seris, pur charyté,
> Redis this caral reuerently,
> Fore I mad hit with wepyng eye,
> Your broder, Jon, the blynd Awdlay.

The third, on St. Anne, could be sung either on her day, 26 July, or at Christmas.

Many of the religious didactic carols take as their theme the transitoriness of life, and the day of doom to come:

> This word,[1] lordynges, is but a farye;[2]
> It faryt ryght as a neysche[3] weye,
> That now is wet and now is dreye,
> Forsothe, serteyn, as I you say.
>
> Now is joye and now is blys;
> Now is balle[4] and bitternesse;
> Now it is, and now it nys;[5]
> Thus pasyt this word away.

But they shade off into others in which the specifically Christian note is not apparent, and which might perhaps best be called carols of the Wisdom of Life. Some of them are frankly pagan in their outlook. Jak Rekles sings,

> Me thynk this word is wonder wery
> And fadyth as the brymbyll[6] bery;
> Thefor Y wyll note but be mery;
> How long Y xall Y cannot sey.

---

[1] word (world).   [2] farye, 'illusion'.   [3] neysche (nesh), 'moist'.
[4] balle (bale), 'evil'.   [5] nys, 'is not'.   [6] brymbyll (bramble).

Certainly the fifteenth century had its full share of the changes
and chances of fortune. There are carols on the untrustworthi-
ness of friends, and on the full and empty purse:

> Peny is an hardy knyght;
> Peny is mekyl[1] of myght;
> Peny, of wrong he makyt ryght
> In euery cuntre qwer he goo.

Others are in lighter vein, on the baselard or dagger, on
hunting, on bad singers, on the troubles of a schoolboy, on the
merits and dangers of ale and wine, and the demerits of women,
especially wives. There are amorous carols, too, which sometimes become wanton. One might suppose them to hark back
to the primitive topics of the dancing throng. But men in all
ages have been amorous, and sometimes wanton. How far the
temper of the Christmas festival was comprehensive enough to
accept all these things one cannot say. It had, of course, its
own secular carols, on the rejection of Advent, on the contest
between Holly and Ivy for the 'maystry' of the hall, or on the
Boar's Head, perhaps sung in procession, as it is now at Queen's
College, Oxford, while the traditional dish was borne up to the
high table. But it is difficult to think that some very realistic
complaints of seduced maidens were delivered in public. There
are political carols, too, chiefly on Agincourt or on the red and
white and Tudor roses. They recall the dancing songs at the
reception of thirteenth- and fourteenth-century sovereigns. But
again, a note in a manuscript tells us that one on Agincourt was
'for Crystynmesse'. Another, on the restoration of Henry VI in
1470, hails his supporters, including 'Wylekyn', who is the Earl
of Warwick, and a 'Lorde Fueryn', who may conceivably be
Thomas Fauconberg, although he was not in fact a lord. The
author, one Jonys, may be Edward Jones, a Gentleman of the
Chapel, who died in 1512.

Carols have come down to us from very heterogeneous
sources. Many are scattered in small numbers over manuscripts of a miscellaneous character, often composite or in more
than one hand, where they find a place among matter of various
kinds, which may be historical, scientific, or domestic, as well
as religious. A few have been gleaned by modern researchers
from surviving usage. There is rarely any indication of the

---

[1] mekyl (mickle), 'much'.

authorship of a carol so found. A name, preceded by a *Per me* or *Scriptum per*, may be merely that of a scribe. There are, however, ten more important collections. Probably the earliest of these is that preserved in a manuscript (Sloane MS. 2593) at the British Museum. It is unfortunately very imperfect, lacking the first forty-eight of its original eighty-four folios. The hand is ascribed to the first half of the fifteenth century. At some time one Johannes Bardel seems to have been the owner. Here are fifty-seven English carols, nineteen of which also exist in other versions. There are also seventeen other poems, English and Latin, of various types. We cannot, of course, assume a single authorship, but the collection may well represent the repertory of a single singer. The earliest editor, Thomas Wright, thought that he was a minstrel, but the free use of Latin does not suggest a minstrel of the normal type. Obviously the poems are not necessarily all by one hand or of one date. There are errors which suggest transcription. One of the Latin poems is adapted from a Goliardic song, inherited from the twelfth century. But I do not think that there is any reason for putting any of the English carols before the fifteenth. One of them, indeed, refers to 'the pestelens tweye', which are probably those of 1348–9 and 1361–2, and to a great wind of 1362, which damaged the tolbooth and Carmelite friary of Lynn in Norfolk, but makes no mention of a third pestilence of 1369. But local memories of such events live so long that this does not seem to me convincing evidence of contemporary authorship. It is likely enough, however, that the writer of this carol came from East Anglia. Another in the collection is on the unusual theme of St. Edmund, the patron saint of Bury St. Edmunds. Most of the carol types already indicated are well represented in the Sloane collection. There are many examples definitely for Christmas use, many on the Virgin, many of a religious-didactic character, a few of wisdom of life or frankly secular. The items which are not English carols include, besides the Goliardic lines, two Latin carols for Christmas, seven religious lyrics, four secular lyrics, and two narrative poems. These last, one on St. Stephen, the other on a greenwood theme, have sometimes been regarded as early ballads, and will be considered in another chapter. One of the secular lyrics is more suitable for the entertainment of the butler's pantry than for that of the open hall. Perhaps this is also true of one of the English carols. On the other hand, two of the non-carol religious pieces reach the highest level of lyrical beauty of which medieval poetry was capable, and shall be given here. The first is on the Virgin:

> I syng of a mayden
> that is makeles,[1]
> Kyng of alle kynges
> to here sone che ches.[2]
>
> He cam also stylle
> ther his moder was,
> As dew in Aprylle
> that fallyt on the gras.
>
> He cam also stylle
> to his moderes bowr,
> As dew in Aprille
> that fallyt on the flour.
>
> He cam also stylle
> ther his moder lay,
> As dew in Aprille
> that fallyt on the spray.
>
> Moder & maydyn
> was neuer non but che.
> Wel may swych a lady
> Godés moder be.

The second is on a more unusual theme:

> Adam lay i-bowndyn,
> bowndyn in a bond,
> Fowre thowsand wynter
> thowt he not to long;
> And al was for an appil,
> an appil that he tok,
> As clerkis fyndyn wretyn
> in here book.
> Ne haddé the appil také ben,
> the appil taken ben,
> Ne haddé never our lady
> a ben hevené qwen.
> Blyssid be the tyme
> that appil také was,
> Therfore we mown syngyn
> *Deo gracias.*

---

[1] makeles (matchless).   [2] ches (chose).

A third, again on the Virgin, is in part based on a Latin hymn by Venantius Fortunatus. There are in all eight poems, religious and secular, of similar metrical types in the manuscript. Possibly they represent a common authorship. On the other hand, Dr. Greg has found the first and last lines of 'I syng of a mayden' in a thirteenth-century collection, and therefore the compiler of the Sloane MS. may have inherited it, perhaps not quite in its present form, from a predecessor.

A second manuscript (Bodleian Douce MS. 302), also of the first half of the fifteenth century, is of a different type from the Sloane MS. It contains the poems of a single author, the John Awdelay to whom reference has already been made. In the manuscript his name also appears as Audelay, Awdlay, Audlay and Audley. It is not, however, in his own hand. He seems to have been both blind and deaf during the whole period of composition. Nine poems are missing from the beginning of the collection. Fifty-five remain. Most of these appear to have been dictated to a single scribe, and to have been read over to the author by another, to whom he indicated many corrections, together with the full text of a final piece. Originally he meant to end his work with what is now the eighteenth poem, which begins with the line 'Here I conclud al my makyng', and to give to the whole collection the alternative titles of 'The Cownsel of Conseans' or 'The Ladder of Heuen'. He is writing

> As I lay seke in my langure
> In an abbay here be west.

Then comes a colophon, dated in 1426, in which he describes himself as a chaplain, as *secus*[1] *et surdus in sua visitacione*, and as writing for the example of others in the monastery of Haghmon. He lived, however, to write thirty-seven more poems, of which twenty-six are carols. The last poem is followed by another colophon, which contains a warning, unfortunately disregarded, to 'kutt owte noo leef', for it will be sacrilege, a promise that anyone who asks shall have a copy, and a further autobiographical note that

> The furst prest to the Lord Strange he was,
> Of thys chauntré her in this place,
> That made this bok by Goddus grace.

Haghmon was an Augustinian monastery near Shrewsbury,

---

[1] secus (caecus), 'blind'.

and Awdelay's patron was probably Richard Lord Strange of Knockyn, also in Shropshire, who died about 1449. A chantry in Haghmon was granted to his son John Lord Strange in 1476, but that occupied by Awdelay must have been an earlier foundation. Of his life before he entered Haghmon we know nothing. He more than once describes himself as sinful and chastised for his living. I once suggested that he might have had a Goliardic youth, but I now doubt whether the phrases are more than those of conventional piety. At Shrewsbury were the relics of St. Winifred, to whom Awdelay had a special devotion. Two poems in his early group are to her. One of these describes her martyrdom, the setting up of her shrine in the town, and the miracles done there.

> Glad mai be al Schrosbere
> To do reuerens to that lady;
> Thai secke here grace and here mercy
> On pilgrymage ther euere Fryday.

The dialect of the poems is north-west Midland, with some traces of contamination both by standard and by northern English. There is often some use of sporadic alliteration, and two of the concluding poems have a full, although rather irregular, alliterative stress. Awdelay's editor finds in these many words which do not occur elsewhere in the manuscript, and is inclined to doubt his authorship. But he is throughout an experimenter in metre, using stanza-forms which sometimes extend to eleven or thirteen lines. On the other hand, he is inclined to repeat himself, both in the choice of themes and in actual phrasing. Some of the early poems are very long and read like versified sermons, with appeals to 'lordis' or 'seris' or those 'that beth present'. One he himself calls a sermon, but another a treatise. But there are also many 'Hails' and orisons, and some narrative poems of a legendary type. The main group of twenty-five carols includes five of which versions are found elsewhere, two of them in the Sloane MS. Dr. Greene doubts Awdelay's authorship of four of these, on grounds which, where he states them, seem to me rather hypercritical. Awdelay's carol metres are simpler than those which he uses in the non-carol poems, and the singing note is often well held, especially in a series for the six principal feast days from the Nativity to the Circumcision. But later he seems to forget

the singing throng, introducing his didactic personality unduly,
with an 'I' in the refrain, where it is clearly unsuitable to a
chorus. And he even becomes autobiographical. A carol, of
which the refrain is *Timor mortis conturbat me*, is wholly concerned
with the physical troubles of 'blynd Awdlay'. It has an explana-
tory stanza:

> As I lay seke in my langure,
> With sorow of hert & teere of ye,
> This caral I made with gret doloure;
> *Passio Christi conforta me.*

Another, on the young Henry VI, which we may think of, if we
like, as written when the boy king kept his first solemn Christmas in 1426, has again a personal ending.

> I pray ʒoue, seris, of ʒour gentré,
> Syng this carol reuerently,
> For hit is mad of Kyng Herré;
>   Gret ned fore him we han to pray.
>
> ʒif he fare wele, wele schul we be,
> Or ellis we may be ful sore,
> Fore him schal wepe moné an e;[1]
>   Thus prophesis the blynd Awdlay.

And a third, on St. Francis, which is admittedly for readers,
ends with yet one more allusion to the author's infirmity. On
the whole I feel that Awdelay is rather remote from the main
tradition of carol development.

One other manuscript (Trin. Coll. Camb. O.3.58) is of a
different type from the two already described. It can be dated
fairly closely as of 1415–22, since it contains a carol on the battle
of Agincourt, and Henry V is still alive. But it is a vellum roll,
in which the carols are set to music, not for a soloist and choir,
but as part-songs for two or in some cases three voices. It is
possible that the composer may be John Dunstable, who died
in 1453, and whose influence on the development of music was
considerable abroad, although it seems to have faded away in
England during the Wars of the Roses. Here are thirteen
carols, most of them clearly intended for use at the Christmas
feasts. They must have had a considerable circulation, since

----

[1] e (eye).

nearly all of them also appear elsewhere, including one in the Sloane MS., one in Awdelay's, which Dr. Greene thinks too spirited for him, and no less than six in another manuscript (Bodleian Arch. Selden B 26). This last has been variously dated, but it includes the Agincourt carol, with a slightly altered reference to Henry V, again apparently as alive. The dialect is southern. There are twenty-three carols, of which two are in the Sloane MS. and seven in collections of later date. Here, too, the carols, again mostly for Christmas, are set as part-songs. There is also a musical composition by one Lionel Power, probably a follower of Dunstable. Other work by him is traceable both in England and abroad. And there are eight English songs, religious and secular, of a non-carol type. One of these is of high literary merit, and shall be given here:

> The merthe of alle this londe
> Maketh the gode husbonde,
> With erynge[1] of his plowe.
> I-blessyd be Cristés sonde,[2]
> That hath vs sent in honde
> Merthe & ioye y-nowe.
>
> The plowe goth mony a gate,[3]
> Bothe erly & eke late,
> In wynter in the clay,
> A-boute barly and whete,
> That maketh men to swete;
> God spede the plowe al day.
>
> Browne Morel & Gore
> Drawen the plowe ful sore
> Al in the morwenynge;
> Rewarde hem, ther-fore
> With a shefe or more
> Alle in the evenynge.
>
> Whan men be-gynne to sowe
> Ful wel here corne they knowe
> In the mounthe of May.
> Howe euer Janyuer blowe,
> Whether hye or lowe,
> God spede the plowe all-way.

[1] erynge, 'ploughing'.   [2] sonde, 'message'.   [3] gate, 'way'.

Whan men by-gynneth to wede
The thystle fro the sede
In somer, whan they may,
Gode lete hem wel to spede
& longe gode lyfe to lede,
All that for plowe-men pray.

From the second half of the fifteenth century come three
collections. One (Bodleian MS. Eng. Poet. E. 1) is a long one.
It is written in two hands, one of which has also contributed a
tune, which is misplaced in the manuscript. Here are seventy-
five English and Latin songs, of which sixty-three are English
carols. Twenty-eight of these are also found elsewhere, two of
them in the Selden MS. and no less than six in the early Sloane
MS. The series is a fully representative one, covering all the
main types. Carols of Holly and Ivy and of the Boar's Head
are among those for Christmas. The second collection (Camb.
St. John's Coll. S. 54) is a short one, with only sixteen complete
carols, fragments of three others, mutilated through injury to
the manuscript, and two non-carol religious poems. One of the
carols and one of the non-carol poems are also in the Sloane
MS., but there the latter has been converted into a carol by
prefixing a refrain. Three other carols are elsewhere. The
manuscript itself is an interesting one. It has a vellum cover,
which has been folded over in a way suggesting use as a
pocket-book. It may contain the repertory of a travelling
singer. The third collection can be more precisely dated. It
forms the greater part of the poetical work (Camb. Univ. MS.
Ee.1.12) of James Ryman, a Franciscan, whose house is unknown. Most of the manuscript is written by a scribe, and
apparently corrected by the author. Then comes a colophon,
dated in 1492, 'Explicit liber ympnorum et canticorum, quem
composuit frater Iacobus Ryman ordinis minorum'. The rest
may be in Ryman's own hand. But a humorous *fabliau* on *The
Fox*, ultimately derived from the twelfth-century *Roman de
Renard*, which also lies behind Chaucer's *Nonnes Preestes Tale*,
can hardly be his. Perhaps he recorded it with a view to turning
the Fox into an analogue of Satan. In all, the collection contains 166 English and Latin items, of which 119 are English
carols. As Dr. Greene points out, these amount to a quarter
of the total number of carols surviving from dates up to 1550.
He is inclined to doubt Ryman's authorship of two of them,

one because it is a rather secular Farewell to Advent, the other apparently because it looks like a later addition at the beginning of the manuscript. Beyond the 1492 of the colophon, there is little evidence as to the date of the carols. Possibly a non-carol poem, on the death of Henry VI, but with no hint of a murder, may be as early as 1471. Ryman is not an inspired writer, and his contribution lays a heavy weight on Dr. Greene's pages. He has some skill in interweaving English and Latin, both in stanza and refrain. Occasionally he exhorts to merriment in song, but he gives little encouragement to it. His approach is devotional rather than didactic. Twenty of his carols are more or less paraphrases of the *Te Deum*. Most of the others are in honour of the Virgin. And he repeats himself indefinitely. Again and again Christ is the 'King of Bliss' or Daniel's 'stone cut of the hill'. The Virgin is the 'Queen of Bliss', the 'maiden mild', the 'flower of virginity', the 'strong Judith', the 'meek Hester', the 'fleece of Gideon', the 'throne of Solomon', the 'closed gate of Ezechiel', the 'florigerate yard of Jesse', the 'lantern of eternal light'. The Incarnation is the sunbeam passing through the glass. It is unlikely that Ryman's work had much circulation. One carol is found in a print of 1550, and there are three other versions of the one which Dr. Greene thinks an addition. This perhaps rather suggests that he may be right. But it contains one of Ryman's favourite phrases.

With the sixteenth century we come upon a collection of a different type (Oxford, Balliol College MS. 354). It forms part of a household book, which also contains much other poetry, religious and secular, narrative and lyric, together with prose treatises, recipes and prescriptions, puzzles and riddles, and business notes, often upon London customs. The compiler was fairly literate, writing Latin and French as well as English. One entry records his name, 'Iste liber pertineth Rycardo Hill seruant with Mr. Wynger alderman of London'. John Wingar, a grocer, became Mayor of London in 1504 and died in 1505. Other notes tell us that Richard Hill was born at Hillend in Langley, part of the parish of Hitchin, Herts., and married Margaret, daughter of Henry Wingar, a haberdasher in the London parish of Bow, and had children from 1518 to 1526, and that he was a merchant adventurer, who visited Antwerp and became a freeman of London in 1511. A mayoral calendar is rather fully annotated from about 1497 onwards, and ends with the election of a mayor in 1535. Probably Hill died shortly afterwards. Among the poems he preserved is one by Dunbar, which he described as 'a litill balet made by London, made at Mr. Shawes table by a Skote'. Possibly he heard it when Sir John Shaw was mayor in 1501–2. At any rate we get a fair range of dating for the compilation of the manuscript. It is now rather disordered, but all the carols appear to be in the same hand, presumably Hill's own. There are seventy-eight of them. Thirty-six are found elsewhere, including fifteen in the Bodleian Eng. Poet. MS., and rather unexpectedly as many as seven in the early-fifteenth-century Sloane MS. Evidently many carols had a long life. One of Hill's appears as late as 1591, adapted by Thomas Mawdycke to be used as a song in the Coventry miracle play of the *Pastores*. Other manuscripts from the first half of the sixteenth century contain carols, which are of the part-song type with tunes by learned composers. The most important of these is in the British Museum (Addl. 5665). Here are thirty-six carols, of which seven are elsewhere, mostly in the Selden MS. Headings indicate that most of them were intended for the Christmas season. The rest might be given *ad placitum*. The composers include a J. D. who may be the early John Dunstable, a John Trouluffe, and a Richard Smert, curiously described in one entry as 'Smert hared de Plymptre'. Plumtree is a town in Nottinghamshire. 'Hared' might possibly be a form of 'Herald'. In another British Museum MS. (Addl. 5465), once owned by Robert Fairfax, a musician who died in 1529, are nine carols, with settings by Gilbert Banastir, Sir Thomas Phelyppis, Edmund Turges, and one Sheringham. In another (Addl. 31922) are five, of which two, one religious and one secular, are ascribed in headings to King Henry VIII, and two others are signed respectively by 'Cornyssh' and 'D. Cooper'. In yet another (Royal Appendix 58) are three, unsigned. The composers may or may not be responsible for the arrangement of the words as well as the tunes. Carols now began to get into print. In 1520 John Dorne of Oxford was selling 'Kesmes corals' at 1*d*. on single leaves and 2*d*. on double ones. Fragments of some early editions in volume form have come down to us. They are generally entitled *Christmas Carolles*. The earliest is from the press of Wynkyn de Worde in 1520. It has two carols. The next, also from him, is of 1530. It was in parts, of which the

Bassus survives. Here are twenty 'songes'. Dr. Greene says that only the music of one of them indicates that it was sung in carol fashion. I do not know that he means more than that the others are set as part-songs. Certainly the words of ten of them are arranged in stanza and refrain. The composers, where named, are Thomas Ashwell, Robert Cowper, John Gwynneth, Robert Jones, Richard Pygot, and John Taverner. Another fragment, perhaps from the press of William Copland, about 1550, has one carol and bits of three others. More important is a complete collection issued, also about 1550, by Richard Kele. It has twenty-one carols, of which seven are found elsewhere. One of these is again in the Sloane MS., written more than a century before Kele's time.

In considering carols which are found in more than one manuscript a distinction must be drawn. There are seventy-eight of them in all, out of Dr. Greene's total number of 474. And in many cases there is much textual divergence. Often this is merely a matter of scribal inaccuracy. A medieval copyist was rarely faithful to his original. He might easily alter the wording of a line or omit one or more stanzas. But often the variation goes far beyond this. What is evidently the same carol in substance may appear with different refrains. The order of the stanzas, as well as their number, may be different. The wording of one version, while obviously indebted to that of another, may be little more than a paraphrase of it. Taken together, these features, which are particularly notable in later carols comparable with those in the Sloane MS., clearly point to a practice of deliberate recasting. Dr. Greene recognizes this, and would find an explanation, rightly I think, in 'the activity of some professional class, literate if not learned'. But I become doubtful when he goes on to identify this class as still in the fifteenth century the mendicant order of Franciscans, which had admittedly been mainly responsible for the earlier development of the carol form in the fourteenth. But the character of Franciscanism had much degenerated in the interval, and Dr. Greene can only point in defence of his thesis to James Ryman, whose carols are just those which seem never or hardly ever to have got abroad. I am myself inclined to attach more importance to the activity of musical composers, whose instinct would certainly be to adapt existing words to their tunes, rather than their tunes to the words they found. The history of English music begins with the development of choirs in royal chapels, on the model of that in the papal chapel at Rome. I have noted the early names of John Dunstable, Lionel Power, and Edward Jones. These are only the forerunners of a very large number in Tudor days. Of some of the latter we know a little more than their bare names. Many first emerge in the sixteenth century, including nine of the thirteen recorded as setting carols. But the other four, Gilbert Banastir, Thomas Ashwell, Robert Jones, and John or William Cornish, had already begun their work before the end of the fifteenth. Many composers were Masters of Song or singing Gentlemen of the Chapel Royal. We have a complete list of the Masters for a considerable period, John Plummer (1444–55), possibly the same as a John Pyamour, who was impressing boy choristers for the King in 1420, Henry Abyngdon (1455–78), Gilbert Banastir (1478–86), apparently supplanted by John Melyonek under Richard III in 1484, Lawrence Squier (1486–93), William Newark (1493–1509), and William Cornish (1509–23). This Cornish, who appears in Addl. MS. 5465 as William Cornish Junior, was probably a son of the William or John Cornish who set a carol for the Queen in 1493. But composers are also to be found elsewhere than at court. The elder William Cornish was Master of the Song-School at Westminster Abbey in 1479–80. Others are recorded as belonging to the choirs of Lincoln, St. Albans, and Wells, to those of Magdalen College and Cardinal College in Oxford, and to those of the private chapels of the Countess of Richmond and Derby and Cardinal Wolsey. On the other hand, the fourteenth-century accounts of the great Priory of Durham suggest that it then largely depended for Christmas entertainment on the services of minstrels. How far, if ever, minstrels came to sing carols we do not know. An early editor of the Sloane MS. suggested that it contained the repertory of a minstrel. Its contents, however, are mainly religious, although, if it belonged to an ecclesiastic, some of them show that he must have been rather a Friar Tuck. The folded wallet of the St. John's College MS. might also, I suppose, be claimed as a minstrel's.

Most of the carol writers seem to be faithful to a traditional stanza form. Dr. Greene calculates that it is found in 188 out of the 474 examples in his collection. The body of it consists normally of a four-stressed triplet on a single rhyme, followed by a tail-line, which rhymes not with the triplet but with the refrain. There are minor variations. The stressing is often irregular. Scribal errors may have contributed to this. In the early Sloane MS. assonance sometimes takes the place of rhyme. The body of the stanza may have two or four lines, instead of three. The tail may be of two lines. The rhyme link with the refrain may be missing. Such a metre may be regarded as a form of *rime couée*. It differs from the *rime couée* found so commonly in miracle plays, because it is not bipartite. The refrain takes the place of the second part. The bipartite form, however, does occasionally appear, as one of many variant stanza arrangements, most of which seem to be experimental and do not recur. But the commonest type, next to the triplet, is the cross-rhymed quatrain, of which there are about eighty examples. In these a rhyme-link between stanza and refrain is less frequently found. Where it is, it may be introduced by one or two tail-lines. Awdelay is fond of the one-line variety and Ryman of the two-line. Stanzas in octaves or in the four-stressed variety or rhyme royal are rare, and for a few of still greater length the part-song composers seem to be responsible. The typical triplet form is closely analogous to that used in the French fourteenth-century dance *caroles* and *chansons de toile*, in the earliest of which assonance also appears instead of rhyme. The same form is found also in some of the Latin *cantiones* or *cantilenae* often introduced into festal church services, and in songs of the *Carmina Burana*.

The refrain is the making of the carol. This is sometimes true, even in a literal sense. Here and there a poem may be found in a non-carol form, which seems elsewhere to have been converted into a carol, mainly by prefixing a refrain. In a sense the refrain is an independent unit. The same refrain may appear in more than one carol, or the same carol may appear with different refrains. In form the refrain is sometimes little more than an outcry, a 'Hey now, now, now', or an 'Alleluia' or 'Nowell', which may be repeated or have the last syllable prolonged. That of one carol introduces the word 'Tyrlo', which may imitate the sound of a shepherd's pipe. More often it takes metrical form, sometimes in a single line, but usually in a couplet, the lines of which normally, but not invariably, rhyme together. Longer refrains are comparatively rare. Some elaborate ones belong to part-song carols. The substance of the

refrain is sometimes, especially in carols for Christmas, little
more than a call to be merry or to sing:

> Make we mery, both more and lasse,[1]
> For now ys the tyme of Crystysmas.

Or,

> Make we myrth
> For Crystés byrth,
> And syng we Yole tyl Candelmes.

Or it bids 'welcum' to Yule or 'Gud day', when the festal
season is over. But normally its function is to set the theme
which the stanzas that follow will develop. In Nativity carols
there is generally a mention of the divine birth, or the election
of the Virgin, or the hope of redemption. There may be one
also of the angelic song or the presence of the Shepherds. A
Twelfth Night carol may similarly refer to the *Magi* or the
guiding star. In devotional carols, including those on saints,
the refrain often takes the form of a prayer or a 'Hail!' Those
on the Virgin are apt to use the special language of her cult.
She is 'a floure of pryce'.

> Ther is no rose of swych vertu,
> As is the rose that bare Jhesu.

Or they use the lines of her antiphons, 'Aue, regina celorum',
or 'Regina celi, letare', or 'Salve, regina, mater misericordie',
or 'Alma redemptoris mater'. In didactic or satirical carols the
refrain often sums up the counsel in a brief phrase, which may
be of proverbial origin:

> An old old sawe hath be fownd trewe:
> 'Cast not away thyn old for newe'.

Or,

> Bewar, sqwyer, yeman and page,
> For seruyse is non erytage.

Carols on death may have the grim line:

> Timor mortis conturbat me.

Strictly speaking, of course, the singers should refer to themselves in the singular during the stanzas and in the plural in the
refrain, at any rate outside part-song. But this principle is not
always observed. The carols of the St. John's MS., in particular,
several times have an 'I' in the refrain. The well-known
Oxford Boar's Head Carol, with the refrain:

> Caput apri defero,
> Reddens laudes domino

[1] lasse (less).

sung while the head is carried in procession up the hall, is probably of late origin.

The element of repetition in carols is not confined to the initial choric refrain. Sometimes each stanza begins with a phrase, which may be varied as the theme develops, but includes words taken from the initial refrain itself. Some of the satirical carols on women and ale furnish particularly good examples of this iterative manner. A more elaborate device consists of a definite secondary refrain, which ends the stanzas and links them together, giving a unity to the whole poem. We may call it the inner refrain. It is particularly common in carols of the triplet type. I find it in about two-thirds of the 288 examples in Dr. Greene's collection. Here it falls upon the tail-line, which either repeats, sometimes with a slight variation, a line, usually the second line if there are two, from the initial refrain, or less frequently only adopts two or three words from it. The inner refrain, so formed, may again be a little varied in the successive stanzas, in order to fit it into the logic of their phrasing. In quatrain carols an inner refrain may similarly fall upon the tail-line, if there is one, or if not, upon the last line of the stanza itself. Sometimes, however, the stanzas of a carol may have an inner refrain of their own, which is not derived from the initial refrain. Awdelay, the preacher, uses the device in a characteristic way to emphasize the lesson which the chorus will repeat, with such introductory phrases as 'Herefore I say', 'Leve ye me', 'I say the so', 'I cownsel the'. Elsewhere he counters a 'Timor mortis conturbat me' of the initial refrain with a 'Passio Christi conforta me' as an inner refrain. And in one carol, which Dr. Greene thinks too good to be certainly his, he makes an effective use of the antithesis between question and response. It begins

What tythyngis bryngst vs, messangere,
Of Cristis borth this New Eris Day?
A babe is borne of hye natewre,
    A Prynce of Pese that euer schal be;
Off heuen and erthe he hath the cewre;[1]
    Hys lordchip is eterneté.
      Seche wonder tythyngis ye may here:
      That God and mon is hon[2] in fere,[3]
      Hour syn had mad bot fyndis pray.[4]

[1] cewre (cure).   [2] hon (one).   [3] in fere, 'together'.
[4] 'Our sin had made us only the prey of fiends.'

A semle[1] selcouth[2] hit is to se:
The burd,[3] that had this barne iborne,
This child conseyuyd in he degré,
And maydyn is as was beforne.
Seche wondur tydyngus ye mow here:
That maydon and modur ys won[4] yfere[5]
And lady ys of hye aray.

Here the tail-lines added to the quatrain are more numerous than is usual, and so it is in some other carols, where possibly the activity of musical composers may be responsible.

Many carols are macaronic, and in them Latin and English lines and phrases are often combined with a syntactical accuracy which indicates a considerable degree of literacy among the writers. Dr. Greene finds Latin used in 202 of his 474 carols. They are naturally, as a rule, of religious or didactic types. But even a lay-minded writer may use Latin. Kele has a ribald carol, with the refrain

Inducas, inducas
In temptationibus.

The Latin is, of course, not classical but accentual. Most of it comes from the hymns of the Office, which are generally, like the carols themselves, in stanzaic form, with four-stressed lines. Dr. Greene enumerates sixty hymns which are so drawn upon. As a rule, only single lines are taken. Carols based throughout upon a particular hymn are rare. Some use is also made of the less metrical 'sequences' or 'proses', which accompany the Mass, of the 'antiphons' sung before and after Psalms, and much of those specially devoted to the honour of the Virgin. A famous prose, known as the *Laetabundus*, attributed to St. Bernard of Clairvaux in the thirteenth century, is traceable through several non-carol English versions and finally contributes lines to two carols. On the other hand, the phrase *Veni coronaberis*, which is used in one carol of the Virgin, and rather oddly in another on the Christmas Ivy, seems to be adapted directly from the *Veni de Libano* of the *Song of Solomon*.

One other theme, probably also of liturgical origin, has considerably affected English medieval poetry, both in its non-carol and its carol forms. It is that of the *planctus*, or lament. The history of this is rather obscure. It starts from the passage

---

[1] semle (seemly). [2] selcouth, 'seldom known', 'wonder'.
[3] burd, 'maid'. [4] won (one). [5] yfere, 'together'.

in St. John's Gospel, in which Christ from the Cross commends
His mother to the care of John. Here neither Mary nor John
speaks. The grief of Mary is dwelt upon in the *Meditationes* of
St. Augustine, and elaborated in the *Gospel of Nicodemus*, both
of which come from the fifth century. Words by Mary come
into a liturgical response of the ninth or tenth century, and with
the growth of her cult in the twelfth century emerge a number
of versified laments by her, among which the most important
are the *Planctus Ante Nescia*, which is in monologue addressed
in part to Christ and in part to the women of Jerusalem, and the
Flete, fideles animae, which adds an appeal to John. These and
other laments, in some of which Christ himself speaks not to
Mary or John but to *Homo*, were used liturgically in the Good
Friday services, where they became attached to the ceremony
of the Adoration of the Cross. It is possible that they were an
important factor in the development of the Passion play. Much
of the early-fifteenth-century *Burial of Christ*, which seems to
have been originally written as a 'treyte or meditatione' and
then converted into a drama, is in substance a lament, at first
between Joseph of Arimathea and the three Marys other than
the Virgin, and later between the Virgin herself and John. It
was written, according to the prologue, 'Off the sorow of Mary
sumwhat to know'. In one passage the Virgin uses the refrain
'Who cannot wepe, come lerne at me'. This is, of course, a late
piece, and the reflex action of lyrical poetry has doubtless
affected it. The lament, or *planctus*, in the strict sense of the
term, was not the only element in the Good Friday services
which laid stress on the physical sufferings of Christ upon the
Cross. The Adoration of the Cross began with the chant
'Popule meus, quid feci tibi?', and this was followed, as the Cross
was raised, with the words 'Quid ultra debui facere tibi, et non
feci?' Here the assumed speaker is again not the Virgin, but
Christ himself. Hymns developed the theme, a *Stabat iuxta
Christi crucem*, and in the thirteenth century a *Stabat mater
dolorosa* by the Franciscan Jacopone da Todi. Literary writings
in Latin also contributed, such as the *Meditationes* attributed to
Bernard of Clairvaux and those of Cardinal Bonaventura of
Padua. And in English non-carol poems from the thirteenth
to the fifteenth century it is widespread. Four types may be
distinguished. There are laments from the Cross, and laments
by the Virgin, both sometimes in dialogue. There are appeals

by the Saviour from the Cross to man, and there are medita-
tions on the crucifixion by the writers in person. Occasionally
these take the form of a direct address to Christ himself or to the
Virgin. But in all variations the physical sufferings which accompanied the Redemption are dwelt upon. There may be an element
of narrative:

> Heʒe vpon a doune,
>   Ther al folk hit se may,
> A mile from the toune,
>   Aboute the midday,
> The rode is vp arered:
> His frendes aren afered,
>   Ant clyngeth[1] so the clay;
> The rode stond in stone,
> Marie stont hire one,[2]
>   Ant seith 'weylaway!'

I have already noted two examples of the theme among the
refrain poems of the fourteenth century, which were the pre-
cursors of the carols, one by William Herebert and one in the
collection of John de Grimestone. It is dear, too, to Richard
Rolle and his followers. Naturally it also passed into the carols
themselves. Even among those which are primarily jubilations
over the event of the Nativity it may emerge. One, which
begins with the refrain 'Hey, now, now, now' and 'This good
tym of Crystmas', has it at great length in a narrative form.
Awdelay uses it, more appropriately, in a dialogue for St. John's
Day. And of course it dominates the special group, sixteen in
number, which Dr. Greene calls 'Carols of the Passion'. Here
we have it both in monologue and in dialogue. Two related
carols, one of which occurs in the Sloane MS. and in two other
very variant versions, use in refrain or stanza a couplet based
upon one in a non-carol poem preserved by Grimestone:

> Mary moder, cum and se:
> Thy Sone is naylyd on a tre.

Another uses, as an inner refrain, a line which I have already
noted as found in a dramatic Burial and Resurrection, 'Who
cannot wepe, come lerne at me'.

> Sodenly afraide,
>   Half wakyng, half slepyng,
> And gretly dismayde,
>   A wooman sate wepyng.

---

[1] clyngeth, 'are shrunk up'.   [2] hire one (hereon).

With fauoure in her face ferr passyng my reason,
And of hir sore weepyng this was the encheson:[1]
Hir Soon in hir lap lay, she seid, slayne by treason.
Yif wepyng myght ripe bee, it seemyd than in season.
> 'Jhesu!' so she sobbid;
> So hir Soon was bobbid,[2]
> And of his life robbid,
> Saying thies wordes, as I say thee:
> 'Who cannot wepe, come lerne at me'.

I do not know why Dr. Greene does not treat as a carol a very similar poem, rather doubtfully ascribed to John Skelton. It has the same stanza-form, and uses the same words in the line

> 'So rubbid, so bobbid, so rufulle, so red'.

And it has a refrain,

> Wofully araide,[3]
> My blode, man, ffor the ran,
> Hit may not be naide,[4]
> My body blo[5] and wanne,
> Wofully araide.

Skelton (c. 1460–1529) was not too late to write a carol. Gilbert Banastir has a rather interesting innovation in a part-song one. The crucifixion, with speeches by Mary and John, comes in a dream. The refrain is sung twice over at the beginning:

> My feerfull dreme neuyr forget can I:
> Methought a maydynys childe causeless shulde dye.

The lament is provided with a rather unusual setting in several of the carols which take the form of lullabies. They owe their origin to the two fourteenth-century lullabies in John de Grimestone's manuscript. There are thirteen of them. Dr. Greene makes them twelve only. But one example, which he seems to regard as a source-poem only, is really an independent one, with a refrain written out, rather unusually, at the end of each stanza. It is an attractive one, based on a common simile for the divine Mother:

> He sayd 'Ba bay';
> Sco sayd, 'Lullay',
> The virgin fresch as ros in May.

The word 'Lullay' or its equivalent occurs in all thirteen, either in the refrain or an inner refrain or in the stanzas. The

---

[1] encheson, 'occasion'.   [2] bobbid, 'beaten'.   [3] araide, 'conditioned'.
[4] naide, 'denied'.   [5] blo, 'livid'.

singing is always by the Virgin to the Child, who Himself miraculously speaks from His Mother's arms or the cradle. In one carol He is rather inappropriately given the refrain:

> Moder, whyt as lyly flowr,
> Yowr lulling lessyth my langour.

But His words usually come in the stanzas, either in monologue or in dialogue with the Mother. Sometimes they are slight, but often they amount to a regular lament, anticipatory of the crucifixion and its torments, in which the Virgin may join. One lament is hers alone, and in another Joseph is also a speaker. Both the stanzas and the refrain tend to be elaborate, and probably the type appealed to the musical composers. But it occurs already in the Sloane MS. The lament from the Child is not the only feature which is rather marked in the lullabies. Often they have also a narrative *Incipit*, in some such form as this:

> This endrys nyght
> I saw a syghth,
>   A mayd a cradyll kepe,
> And euer she song
> And seyd among,
>   'Lullay, my chyld, and slepe'.

The *Incipit*, again, is sometimes treated as a refrain, although its words are obviously not choric. And it may be compared with the long lullaby poem in John de Grimestone's collection, of which the first stanza is:

> Als I lay vpon a nith
> Alone in my longging,[1]
> Me thouthe I sau a wonder sith,[2]
> A maiden child rokking;

and the last is:

> Serteynly, this sithte[2] I say,[3]
> This song I herde singge,
> Als I lay this Yolis Day[4]
> Alone in my longgingge.

But the *Incipit* has a still older ancestry. It goes back, like the carols themselves, to the manner of certain French courtly poems from the twelfth century onwards, the *chansons d'aventure*, in which a *pastourelle*, or other theme with an element of

---

[1] longging, 'longing' or, possibly, 'lodging'.　　[2] sith, sithte (sight).
[3] say (saw).　　[4] Yolis Day (Yule-day).

narrative, has normally a similar brief autobiographical setting, such as:

> L'autre jor me chivachai
> Toz pensis et en esmai[1]
> d'amors qui m'argue.[2]

And the singer finds consolation in the favours of a shepherdess. Similar openings are common in English non-carol poetry, both before and during the fifteenth century. In carols, outside the lullabies, they are less often to be found. This is natural enough, since narrative, although occasionally present in carols, does not dominate them. In a few cases English minstrelsy provides an alternative *Incipit*:

> Lestenytgh, lordynges, both grete and smale;
> I xal you telyn a wonder tale.

What then is to be said of the carols from a literary angle? There are some dull examples, and towards the end some over-elaborated ones. But on the whole they compare favourably, not only with the formal verse of contemporary ecclesiastical writers, but with anything which the sixteenth-century poets have to give before the days of Sir Thomas Wyatt, who indeed borrowed alike their favourite triplet form and their use of refrain. They are of course communal poetry, and lack the personal utterance, in response to life and its problems, which we have come to look upon as fundamental in the deepest lyric. But they are free from the aureate language which infects so much fifteenth-century writing, and the singing note is often well held, especially in the Nativity carols with their jubilant

> Man be merie, as bryd on berie,

although this is occasionally varied by a more pessimistic touch:

> Who wot nowe that ys here
> Where he schall be anoder yere?

> Anoder yere hit may betyde
> This compeny to be full wyde,
> And neuer onodyr here to abyde;
> Cryste may send now sych a yere.

The lightness is often helped by a free use of anapaestic rhythm. The dull Ryman cannot use the anapaest; it is a main cause of his dullness. There are felicitous touches of description in

---

[1] esmai, 'disquiet'.   [2] argue, 'urges'.

narrative passages, which make one think of the early Italian
pictures, full of light and colour:

> It fell vpon high mydnyght:
> The sterres shon both fayre and bright;
> The angelles song with all ther myght.

The Christ-child is beheld 'in a manjour of an ass', or 'betweene
beestys two', or 'in a crache¹ with hey and gras'. The Kings
'off gret noblay' come in:

> To worchepyn hym thei knelyd on kne
> With gold and myr and francincens.

There is much tenderness to the Virgin, the 'maide, brygt of
ble':²

> Gabryell the angell dyde grett
> Mary kneling in her closett.

Or again,

> His ardene³ he dede ful snel⁴
> He sat on kne and seyde 'Aué':
> Mary stod stylle as ony ston.

The writers have seen and heard these things for themselves in
the miracle plays. In the monitory carols there is a touch upon
the humorous aspects of life, as well as upon its fundamental
seriousness:

> Peny is an hardy knyght;
> Peny is mekyl⁵ of myght;
> Peny, of wrong he makyt ryght
> In euery cuntre qwer he goo.

But,

> Thys world is butt a chery-fare.

There is some vigorous homespun:

> Vnder the busch ye shul tempeste
> Abyde tyl hit be ouer goo.

And,

> Trewthe is far and semit hynde⁶
> Good and wykkyt it haght in mynde;
> It faryt as a candele-ende
> That brennyt⁷ fro half a myle.

The deeper reaches of poetic imagination we must not expect.

---

¹ crache, 'cradle'.   ² ble, 'visage'.   ³ ardene, 'errand'.
⁴ snel, 'quickly'.   ⁵ mekyl (mickle), 'much'.
⁶ hynde (hend), 'near'.   ⁷ brennyt, 'burns'.

Two carols, perhaps, stand out, and shall be given here, as the
best which the fifteenth century can do in this way:

> Mery hyt ys in May mornyng,
> Mery wayys for to gone.

> And by a chapell as Y came,
> Mett Y wyhte Jhesu to chyrcheward gone,
> Petur and Pawle, Thomas and Jhon,
> And hys desyplys euerychone.

> Sente Thomas the bellys gane ryng,
> And Sent Collas[1] the Mas gane syng;
> Sente Jhon toke that swete offeryng,
> And by a chapell as Y came.

> Owre Lord offeryd whate he wollde,
> A challes alle off ryche rede gollde,
> Owre Lady the crowne off hyr mowlde;[2]
> The son owte off hyr bosom schone.

> Sent Jorge, that ys Owre Lady knyghte,
> He tende the tapyrys fayre and bryte,
> To myn yghe a semley syghte,
> And by a chapell as Y came.

Dr. Greene thinks that the refrain may be from a secular May-
song, but this we have no means of checking, although it is
likely enough that such songs were still sung and perhaps even
danced in the fifteenth century. But what is the significance
of 'wyhte Jesu'? The *Dictionary* gives 'wyth' as a fourteenth-
century form alike for 'with', 'white', and 'wight' in the sense of
'strong, mighty', and for 'wight' finds both 'wyht' and 'wyhte'
still in northern use during the fifteenth. It would add significance to the line if we accepted 'white'. In that case the carol
might be, exceptionally, one for use at Whitsuntide, which can,
of course, fall in May. The other outstanding carol is even
more obscure:

> Lully, lulley; lully, lulley;
> The fawcon hath born may mak[3] away.

> He bare hym vp, he bare hym down,
> He bare hym into an orchard brown.

> In that orchard ther was an hall,
> That was hangid with purpill and pall.

---

[1] Collas, 'Nicholas'.   [2] mowlde, 'top of head'.   [3] mak, 'mate'.

And in that hall ther was a bede;
Hit was hangid with gold so rede.

And yn that bed ther lythe a knyght,
His wowndes bledyng day and nyght.

By that bedes side her kneleth a may,[1]
And she wepeth both nyght and day.

And by that beddes side ther stondith a ston,
'Corpus Christi' wretyn theron.

This carol is in more than one way abnormal. The 'Lully'
refrain seems inappropriate, since it is not a lullaby. And the
chivalric setting is unusual. Oddly enough, three poems,
evidently related to this carol, have come down to us in
traditional forms. The earliest was reported by James Hogg,
who was collecting material for Scott's *Minstrelsy of the Scottish
Border* in 1802. The others turned up later in Staffordshire and
Derbyshire. All are mainly in couplet form. The Scottish
version has no refrain. In each of the English ones the couplets
are preceded by a quatrain which may be meant for one.
Between these two there is probably some relation, since each
introduces a reference to the Glastonbury thorn, believed to
have blossomed on the night of the Nativity. It has been suggested that the original carol was in some way connected with
the legend of the Grail. This is rather a wild conjecture, since
a Grail poem ought surely to have a Grail vessel in it, and there
is none. Moreover, both the English versions call the maiden
the Virgin Mary. I do not myself see anything in the original
except her grief for her Son, imaginatively rendered. Dr.
Greene draws an inference, which rather puzzles me, that the
carol itself is what he calls 'true folk-song'. He quotes a definition of this by the Folk-Song Society as 'song and melody born
of the people and used by the people as an expression of their
emotions, and (as in the case of historical ballads) for lyrical
narrative', and glosses this for himself as 'song, that is, which
first makes its appearance in the oral tradition of a homogeneous
community without "book-learning" '. With all deference,
here are two definitions, not one, a definition by origin and
another by the accident of discovery. The *Corpus Christi* carol
first makes its appearance not in oral tradition but with other
carols in a sixteenth-century manuscript. That it afterwards
got into oral tradition, and was much modified in the process,

---

[1] may, 'maid'.

is no evidence that it was by origin 'true folk-song'. Later Dr. Greene suggests that two indecent carols, 'because of their homeliness, their directness of speech, and their theme of the betrayed girl, have a strong case for consideration as authentic folk-song'. I am afraid that I see nothing in them but the work of some graceless minstrel. A fuller discussion of folk-song will come better in connexion with its relation to the ballad. So far as the carol is concerned, I think that we can only credit it with the initial practice of choric singing, with certain topics which belong to Yule rather than the Nativity, and with the habit of iteration in the refrain, which not infrequently also affects the stanzas. The iterative manner similarly shows itself in non-carol poetry of a type related to the carol, sometimes with a progressive variation, which perhaps makes 'cumulative' the better epithet. Two examples, from the early Sloane MS., have already been given. A third, secular in character, comes from the same Balliol MS. which preserves so many carols:

I haue xii oxen that be fayre & brown,
   & they go a grasynge down by the town!
      With hay, with howe, with hay!
        Sawyste thou not myn oxen, you litill pretty boy?

I haue xii oxen, & they be ffayre & whight,
   & they go a grasyng down by the dyke,
      With hay, with howe, with hay!
        Sawyste not you myn oxen, you lytyll pretty boy?

I haue xii oxen, & they be ffayre & blak,
   & they go a grasyng down by the lak,
      With hay, with howe, with hay!
        Sawyste not you myn oxen, you lytyll pretty boy?

I haue xii oxen, & they be fayre & rede,
   & they go a grasyng down by ye mede,
      With hay, with howe, with hay!
        Sawiste not you myn oxen, you litill pretty boy?

The carols and the few poems of similar temper associated with them in the manuscripts represent the best which the fifteenth century can do in the way of religious lyric. There is much besides, but it does not compare favourably in literary value with the much slighter output of fourteenth-century work which the chances of survival have preserved to us. The themes remain, on the whole, very much the same. The *Planctus*, or Lament, in the various types of its development, is frequent.

The note of quasi-amorous ecstasy, which Richard Rolle and his followers brought into their depiction of the relations between Christ and the Soul, has disappeared. Perhaps it is hardly to be regretted, in spite of the beauty of the *Quia Amore Langueo*. The tendency to Mariolatry has grown. To the religious feeling of the century the 'Moder and mayde' seemed more accessible than the remoter figures of the divine hierarchy. The approach to her is sometimes affected by the language of courtly poetry. There are pieces in which it is difficult to say whether an earthly or a heavenly mistress is being addressed. In form, the fifteenth century inherits from the fourteenth, besides simpler metres, that of the four-stressed octave or *douzaine*, with a turn-over of rhyme before its last quatrain and a final refrain, which was used so commonly in the earlier Vernon MS. The refrain sums up the significance of the poem, but, more often than in the carols, is varied in wording, to fit the logic or the grammar of the stanzas. It does not suggest singing.

In the main, we are concerned with anonymous poetry. Lydgate, of course, is there, in the more popular manner, which often replaces his Chaucerian one. Of John Awdelay and James Ryman enough has already been said. Richard de Caistre, whose adaptation of an earlier hymn appears in no less than seventeen manuscripts, was a vicar of St. Stephen's, Norwich. A 'Halsham squiere', who is responsible for two short pieces, sometimes found together, has been plausibly identified with a John Halsham who died about 1415. Two or three others have the inscription 'A god whene', which may be an indication of authorship, or may be merely a pious ejaculation. A few bear the names of authors whose identity has eluded research. But the great majority are anonymous. As a body, they read like the work of cloistered ecclesiastics who have mistaken religious fervour or conviction for literary inspiration. They do best when they are simplest, in quatrains or occasionally *rime couée*. But they are fond, not only of the octaves and *douzaines*, but of even more elaborate forms, which they cannot handle with success. Metrically they are incompetent. The anapaest, which gives lightness and variety to four-stressed verse, requires to be used with a discretion which they do not possess. And they have been puzzled by the introduction of rhyme royal and other five-stressed metres, the nature of which they do not fully understand. As a result, you often get an incongruous jumble, containing lines of different length which defer to no principle of rhythm. Puzzled scribes may no doubt have contributed to this result. Sometimes ingenuity is brought in to compensate for the lack of poetic feeling. Lines are so arranged that their initial letters spell out a divine name or the like. Many devotional poems are little more than dull paraphrases of well-known Latin hymns, antiphons, or prayers. But there is a worse feature than this. The writers, as their profession requires, are latinists, and they are bitten with the desire to decorate their homely English with strange terms of Latin origin. The tendency first shows itself in the frequent use for rhymes of words ending in -acion or -ion. But by the middle of the century the 'aureate' language, as it came to be called, was in full career. The better poets were cautious. Lydgate, in one passage, asks for 'auriat licour off Clio'. In another, he expresses regret, perhaps ironically, that 'the aureat beames do not in me shyne', and in a third begs a friend to read his verses:

> And of my penne the traces to correcte,
> Whiche barreyne is of aureat licour.

Skelton, too, is apologetic:

> My wordes vnpullyshed be, nakide and playne,
> Of aureat poems they want ellumynynge.

Hawes fell a ready victim:

> Her redolente wordés of swete influence
> Degouted[1] vapoure most aromatyke,
> And made conversyon of complacence;
> Her depured[2] and her lusty rethoryke
> My courage reformed, that was so lunatyke.

It is to the credit of the early-sixteenth-century humanists that they saved the English tongue from this destruction. But to the ecclesiastical lyrists the 'ink-horn terms' were a God-send. The Redeemer goes to heaven in his 'superlatyve and innosable[3] mycht'. The Virgin is the divine 'solistrice',[4] the 'meyden melleffluus', the 'wife mundificate',[5] the 'flowre of all formosité',[6]

---

[1] Degouted, 'distilled'.   [2] depured, 'purified'.
[3] innosable, 'unknowable'.   [4] solistrice (solicitress).
[5] mundificate, 'cleansed'.   [6] formosité, 'beauty'.

the 'ros intact' with 'no spue of crime coinquinate'.[1] One
singer perhaps outdoes the rest:

> By the spectable splendure of hir fulgent[2] face
> My sprete was rauesched, & in my body sprent,[3]
> Inflamed was my hert with gret solace
> Of the luciant[4] corruscall[5] resplendent.

With such language they eke out their barrenness.

They are sometimes happier in monitory poems, when they
dwell on the instability of life and the imminence of death. The
fifteenth century, with its warfare and revolutions and the break-
down of law between man and man, had hard lessons to teach,
even when it was only viewed through the grates of a monastery.
The refrain poems, in particular, offer much homespun wisdom
in their gnomic endings. Lydgate has the best of them, with
'He hasteth weele that wysely can abyde', and 'Experience
shewith the world is variable', and 'See myche, say lytell, &
lerne to suffer in tyme', and 'Look in thy merour and deeme[6]
noon other wiht',[7] and the more imaginative 'Al stant on
chaung, like a mydsomer rose'. But the others sometimes chime
in with 'This world ys but a wannyté', or 'God send us pacyence
in our old age', or 'Whate euer thowe say, A-vyse the well'.
The *Ubi Sunt* motive, which is as ancient as poetry itself, still
prevails. Thomas of Hales knew it in the thirteenth century.
Lydgate has a long bead-roll in which figure David, Solomon,
Absalom, Julius Caesar, Pyrrhus, Alexander, Nebuchadnezzar,
Sardanapalus, Cicero, Chrysostom, Homer, and Seneca,
together with the 'dozepeers'[8] of Charlemagne and the Nine
Worthies 'with all ther hih bobbaunce'.[9] He has ransacked his
library to advantage. Another, more briefly, contents himself
with three names, to illustrate his moral:

> Thow seis thi sampil euerilk day,
>> And thou tak heid withoutyn les,[10]
> Quhow sone that yowt may pas away,
>> For bald[11] Hector and Achilles
> And Alexander, the prowd in pres,[12]
>> Hes tane thare leif & mony ma,
> That ded hes drawyne one-til his des,[13]
>> *Memor esto nouissima.*

---

[1] coinquinate, 'defiled'. [2] fulgent, 'shining'. [3] sprent, 'sprinkled', 'dispersed'.
[4] luciant, 'shining'.   [5] corruscall, 'glittering'.   [6] deeme, 'judge'.
[7] wiht (wight), 'person'.   [8] dozepeers, 'twelve peers'.
[9] hih bobbaunce, 'high boasting'.   [10] les (lease), 'lying'.   [11] bald (bold).
[12] prowd in pres, 'valiant in the thick of battle'.   [13] des (dais), 'judgement seat'.

The monitory poems shade off into others which can best be described as political. Of these an interesting group is found in a Bodleian MS. (Digby MS. 102). They are in octave stanzas, often with refrains, and their editor thinks that they may be all by a single writer, possibly the abbot or prior of a monastery in the west or south-west Midlands. There is much reference in them to public wrongs through oppressive lords and officers, unjust judges, and flatterers. Monks and priests also came in for criticism, although not from a Lollard standpoint. Some of the topics seem to echo parliamentary discussions during the first two decades of the century. Later poems often deal with specific historical events. Lydgate contributes several. Others may be of more popular origin. The chroniclers tell us that after Agincourt Henry V forbade songs in his praise, but there is at least one on the battle, besides the carol already noted. Two are prayers for the well-being of Henry VI on his voyage to France for his coronation at Paris in 1430. Two are satirical 'in despyte of the Flemynges', on the unsuccessful attempt of the Duke of Burgundy to besiege Calais in 1436. One of these is probably by Lydgate. Others, during the later part of this unhappy reign, mostly come from London sources and reflect the dislike felt in the south of those in power at court. The condemnation of Eleanor Cobham, Duchess of Gloucester, for necromancy, heresy, and treason in 1441 is hailed with the refrain:

All women may be ware by me.

And the murder of the Duke of Suffolk in 1450 with:

For Jack Napes soule *Placebo* and *Dirige*.

The Wars of the Roses yield several Yorkist poems, but only one Lancastrian one, an allegory in *rime couée* on *The Ship of State*. All this is of little account as literature. Rather more interesting is a solemn lament by the soul of Edward IV, with the 'chery fair' and *Ubi sunt* motives in it, and the line,

I have pleyd my pagent & now am I past.

There are satirical pieces. *London Lickpenny*, the authorship of which has been doubtfully attributed to Lydgate, gives a vivid picture of life in the metropolis, which a Kentish ploughman, going to seek justice at Westminster Hall, finds a costly place. The refrain is:

For lack of money I may not speed.

This is echoed in *Syr Peny*, the theme also of a carol:

> Sir Peny over all gettes the gré,[1]
> Both in burgh and in ceté,
>   In castell and in towre.
> Withowten owther spere or schelde,
> Es he the best in frith[2] or felde,
> And stalworthest in stowre.[3]

There is also some sculduddery, which is faithfully garnered by the editors of the *Reliquiae Antiquae*.

Amorous themes are generally left to the courtiers. When more popular writers handle them, they are apt to take the form of palinodes.

> Lord, how shall I me complayn
>   Vnto myn own lady dere,
> For to tell her of all my payn
>   That I fele this tyme of the yere?
> My love, yf that ye will here,
>   Though I can no songis make,
> So this love changith my chere,
>   That whan I slepe, I can not wake.

And again, from another:

> Wel-com be ye whan ye go,
>   & fare-wel whan ye come;
> So faire as ye ther be no mo,
>   As brygth as beré broune.
> I love yow, verraly, at my to,[4]
>   Noune so moch yn al this tounne;
> I am right glad when ye wil go,
>   And sory when ye wil come.

There are, however, two remarkable exceptions. Some youthful votary, familiar with liturgical terms, bethought himself of stringing together his love-ditties and labelling them with terms taken from the opening passages of the *Graduale*. The *Introibo*, *Confiteor*, *Misereatur*, *Officium*, *Kyrie*, *Christe*, *Gloria in Excelsis*, *Oryson*, he called them, and added an *Epystal* in prose. The collection is known as *The Lover's Mass*. His superiors would no doubt have reprehended him as a graceless clerk, but there is nothing ribald about the poems. Technically they are parodies, but in substance they are unusually good lyrics, in a

---

[1] gré, 'goodwill'.
[3] stowre, 'combat'.
[2] frith, 'woodland'.
[4] to (toe).

variety of metres. Perhaps the best movement is the *Gloria in Excelsis.*

Worsshyppe to that lord aboue,
That callyd ys the god of loue,
Pes to hys seruantes euerychon,
Trewe of herte, stable as ston,
        That feythful be.

To hertys trewe of ther corage,
That lyst chaunge for no rage,
But kepe hem in ther hestys[1] stylle,
In all maner wedris ylle,
        Pes, concord and vnyté.

God send hem sone ther desyrs
And reles of ther hoote ffyrs,
That brenneth at her herté sore,
And encresseth more and more,
        This my prayere.

And after wynter wyth his shourys
God send hem counfort of May flourys,
Affter gret wynd and stormys kene,
The glade sonne with bemys shene
        May appere.

To yive hem lyght affter dyrknesse,
Joye eke after hevynesse;
And after dool[2] and ther wepynge,
To here the somer foullys synge,
        God yive grace.

ffor ofte sythe men ha seyn
A ful bryght day after gret reyn,
And tyl the storme be leyd asyde,
The herdys vnder bussh abyde,
        And taketh place.

After also the dirké[3] nyght,
Voyde of the Mone and sterré lyght,
And after nyghtys dool and sorowe,
ffolweth ofte a ful glade morowe
        Of Auenture.

Now lorde, that knowest hertys alle
Off louers that for helpé calle,
On her trouthe of mercy rewe,
Namly on swyche as be trewe,
        Helpe to recure.
        Amen.

[1] hestys, 'vows'.   [2] dool (dole), 'grief'.   [3] dirké (dark).

The other outstanding poem is of course the well-known *Nut-Brown Maid*. This comes to us in two texts. One is from Richard Arnold's *Customs of London* printed by Jan van Doesborch at Antwerp about 1503; the other from that manuscript gathering of Richard Hill's, which also yields so many carols. A third, in the seventeenth-century Percy Folio MS., is too late to be of value. John Dorne of Oxford was selling copies, presumably on single sheets, at 1*d*. each in 1520. The stanza-form of the poem is unusual. It is generally printed as in twelve lines. But two out of every three lines have internal rhymes, and on analysis it resolves itself into an eighteen-line variety of the *rime couée*, so beloved by popular singers, complicated by the use of two distinct *refrains*, which recur with very slight variation. The technical description is thus $aa_2 \, b_3 \, cc_2 \, b_3 \, dd_2 \, b_3 \, ee_2 \, b_3 \, ff_2 \, g_3 \, hh_2 \, g_3$. The poem is an example of the greenwood theme, about which more will have to be said in an account of narrative verse. It is cast on the ancient model of a *conflictus*. At the beginning a man and woman dispute as to woman's faith. She cites the Nut-Brown Maid as an example of it, and they decide to tell the story in dialogue. They take alternate stanzas. His refrain is 'I am a banished man', and hers 'I love but you alone'. He is an outlawed knight, and must live by his bow. It is a shock to her:

> O Lorde, what is
> This worldés blisse,
>   That chaungeth as the mone?
> My somer's day
> In lusty May
>   Is derked before the none.

But she is constant and will follow him. He attempts to dissuade her. She will be thought a wanton. And it is a hard life in the greenwood.

> Yet take good hede
> For ever I drede
>   That ye coude not sustein
> The thorney wayes,
> The depe valeis,
>   The snowe, the frost, the rein,
> The colde, the hete,
> For, drie or wete,

> We must lodge on the plain,
> And, us above,
> Noon other rove[1]
> But a brake bussh or twaine.

She will have to cut her hair by her ear and her kirtle by her knee. All this she will endure, and even, if need be, bear a bow in his defence. Then he tells her, untruly, that he has already another maid in the forest. Even this does not move her. She declares that she will be soft and kind and courteous to her rival. Her trial is over. He is no banished man, but an earl's son. He will take her to Westmorland, which is his heritage, and marry her there. The two speakers join in a final stanza of rather trite piety:

> Here may ye see
> That wimen be
>   In love, meke, kinde, and stable.
> Late never man
> Repreve them than,
>   Or calle them variable;
> But rather prey
> God that we may
>   To them be comfortable,
> Which sometime proveth
> Such as he loveth,
>   If they be charitable.
> For sith men wolde
> That wimen sholde
>   Be meke to them each one,
> Moche more ought they
> To God obey,
>   And serve but Him alone.

It is possible that this piece ought to have been treated under the head of drama, rather than of lyric. It might well have been recited by two minstrels in a baronial hall, as a kind of *estrif*. But in any case it makes a gracious end to the fifteenth century, and to this discussion.

[1] rove (roof)

POPULAR NARRATIVE POETRY AND THE BALLAD

THE distinction between narrative and lyric poetry is a difficult one. It owes its origin, as Professor Ker has pointed out, to Greek usage, in which the epic was a narrative for recitation, while the lyric was sung to music. This was no doubt so at the Panathenaea of the sixth century, although in Homer's own time the heroic lay, which developed into the epic, appears to have been accompanied by the harp. And this seems to be also true of the early Teutonic lays and of those of the Anglo-Saxon *gleómen*, and of many of the *chansons de geste* and *romans d'aventure* which succeeded them. But in course of time *fableors* begin to make their appearance as minstrels of a special type, and these presumably had laid the harp aside and were mere story-tellers. Even in the sixteenth century, however, Sir Philip Sidney could still hear the story of Chevy Chase sung, although the instrument which now accompanied it was the more portable crowd or fiddle rather than the harp. In Scotland and Ireland, however, the harp seems to have long survived. Evidently, as Professor Ker realizes, the sharp Attic dichotomy was not quite applicable to more modern poetry. In the fifteenth century we must be content to take as lyric, besides the carols, for which there is clear evidence of choric singing, most of those pieces which have the more elaborate stanza forms, and in particular those which have refrains; and as narrative those which, even if sometimes sung or at least chanted, primarily describe events and are, as a rule, in simpler couplet or septenar metres. But, as has been noted in the last chapter, there is often much incidental narrative in carols. You cannot always sing about nothing. And on the other hand, even in narrative the reporter does not always exclude his own personality, but may give his story a colouring or atmosphere of triumph, regret, or wonder which makes of it something more than a mere versified chronicle. Sung or not, it becomes a lyrical narrative.

The last flickerings of medieval romance and hagiographical legend find their place in more appropriate chapters of this record. There are a few historical narratives. We do not know the song which, according to Adam of Usk, told of the adventures of Sir John Mortimer in rebellion against Henry IV, early in the century. But the battle of Agincourt in 1415 was celebrated in a long poem, once but now no longer ascribed to Lydgate. It is purely narrative, in spite of the fact that each of its octaves is followed by the refrain

> Wot ye right well that thus it was,
> *Gloria tibi Trinitas.*

Two shortened and much corrupted versions of this also exist. One is partly preserved and partly paraphrased in a text of the *Chronicle of London*. The other was printed as *The Batayll of Egyngecourt* by John Skot about 1530. A similar versified account of the siege of Rouen in 1418 has the *explicit*:

> With owtyn fabylle or fage[1]
> Thys processe made John Page,
> All in raffe[2] and not in ryme
> By cause of space he had no tyme;
> But whenne thys were[3] ys at an ende,
> And he have lyffe and space he wyll hit amende.

This in its turn was utilized, partly in its original form and partly in summary, by the compiler of the prose chronicle known as the *Brut*, which indeed may be largely based on such material. It makes a similar use of a rather vigorous piece on the siege of Calais in 1436. But all this is of course versified journalism rather than poetry.

There are a good many popular narratives which aim at edification or entertainment. As the author of one of them writes:

> Off talys and trifulles many man tellys;
> Summe byn trew and sum byn ellis.

We cannot always be sure that such things, although they come to us in fifteenth-century manuscripts, are not really of earlier origin. It is so with *The Childe of Bristowe*, which has sometimes been ascribed to Lydgate. It has a longish *Incipit* of a minstrel type:

> He that made bothe helle and hevene,
> Man and woman in dayes vij,
>   And alle shal fede and fille,
> He graunte us alle his blessyng,
> More and lasse, bothe olde and yong,
>   That herkeneth and holde hem stille.

---

[1] fage, 'fiction'.   [2] raffe, 'rudeness'.   [3] were (war).

The beste song that ever was made
Ys not worth a lekys[1] blade,
But men wol tende ther-tille;
Therfor y pray yow in this place
Of your talkyng that ye be pes
Yf it be your wille.

Some form of the old *rime couée* stanza, as here used, remains a favourite metre for tales. But a later version of this one is in septenar couplets, with the title of *The Marchand and his Son.* A franklin has acquired great riches by unjust means. He has a son, of whom he would make a lawyer. But the son refuses and becomes apprentice to a merchant. On the point of death, the father can find no one who will become executor to a man noted for his extortions. He sends for his son, who bids him return in three days after his death. He does so return, 'ledd with fendys blake', and 'brenning[2] in flame of fyre'. The son gives him another fortnight:

Here ys a fytt[3] of thys matere; the bettur ys behynde,
Ye schall here how gode Wyllyam to hys fadur was full kynde.

William goes to the merchant who was his master, sells him all his property for a thousand pounds, pays his father's debts, hires a priest and friar to pray for his soul, and makes restitution to those who had been wronged. At the end of the fortnight the father appears again, no longer afire, but still 'black as any pyche, and lothely on to loke'. During his lifetime, he had paid no tithes and given nothing to the church. William was now penniless. He sold himself to his former master as a servant and made further restitution. Last of all appeared an old man on crutches, with a claim for the value of half a quarter of wheat. William could give him nothing but his clothing. And then at last the father appeared 'in grete gladnesse, as bryghte as any sonne'. He was saved at last by the selling, not of William's goods, but of his body. And the merchant gave William his daughter to wife, and he lived happily ever after.

Lythe and lystenyth, gentylmen, that have herde thys songe to ende,
I pray to God, at oure laste day to hevyn that we may wende.

There are other pious tales. In *A Lady that was in Dyspeyre* the wife of a devout husband steals a host and buries it beneath a pear-tree. At Christmas a bishop comes as guest. The pear-

---

[1] lekys (leek's).   [2] brenning, 'burning'.   [3] fytt, 'division of a poem'.

tree blossoms white and red and the lady is alarmed. At the third course ripe pears are falling. A squire breaks a bough, which bleeds. The bishop sets it back to the tree, and it grows again. Men delve to the root, and find 'a blessyd chylde formed in blode and bon' The child blesses the people, but will not look on the lady. She begs the bishop to shrive her. The bishop takes it to the altar, where it becomes bread, and the lady communicates. In *How a Merchand Dyd hys Wyfe Betray* a good wife is neglected for a leman. The merchant, going to sea, borrows money from his wife. She asks him to buy her a pennyworth of wit. He buys jewels for the leman in France. An old man offers him a pennyworth of wit, bidding him go to the leman, and say that he has lost all, and then to his wife. He does so. The leman will have no more to do with him. The wife will work for him. He goes again to the leman, well arrayed. She now gives him riches. He takes them to his wife, and they are reconciled. In *A Father and his Son* there are two brothers, squires of great renown, one of whom was adulterous. Both were slain on the same day. The adulterer went to hell, the other to paradise. The adulterer's son prayed to know where his father was, and was led by one in a white surplice to see him burning in hell and fiends 'with hokis kene' rending his body. The father makes confession to him, bidding him, when a priest, not pray for him, but preach faithfulness in wedlock. The son is then led by an angel to a fair arbour, rather well described, where he sees on a green hill a withered tree dropping red blood, which is that from which Adam plucked the apple, and then a pine-apple of burnished gold, beneath which sits one 'as briȝt as any son beme'. It is his virtuous uncle. In *The Unnatural Daughter* a father debauches his daughter, who kills three children by him. Her mother discovers the intrigue, and the daughter kills her. The father repents and confesses to a priest. He rejects the daughter and plans a pilgrimage. The daughter murders him in his sleep, goes into the country, and lives in lechery. A bishop comes to preach. She enters the church. The bishop sees four fiends leading her by chains. A word he speaks touches her and she weeps. The chains are broken, the fiends flee. She confesses. The bishop bids her wait until his sermon is done, but her heart bursts. The bishop bids the congregation pray to know where her soul is. A voice declares that it is in heaven. The narrator warns sinners not to fall into

wanhope,[1] for they may yet be saved. In *The Knyght and his Wyfe*
devotion to the Virgin meets its reward. The knight spends so
much in honouring her feasts that he becomes poor and can do
so no longer. Ashamed of this, he goes on one feast-day to
hide in a wood until it is over. A fiend, who covets his lady,
tells him where to find gold and bids him return with her on
an appointed day. They reach a chapel, which the wife enters,
while the knight proceeds. The wife falls asleep before an image
of the Virgin, who takes her palfrey and joins the knight in her
likeness. They meet the fiend, who recognizes the Virgin and
cries out. Husband and wife are now safe under heavenly
tutelage. Presumably in these pious tales ecclesiastics have
contributed to the edification and incidentally the entertainment of the folk.

We expect sensationalism in the fifteenth century, but we
do not, perhaps, expect sentimentality towards animals. We
get it, however, in *The Mourning of the Hare*, a lament rather
than strictly a narrative, in which Wat describes the tragedy
of a creature driven from down to dale by hunters who will not
wait to hear their mass, traced by his footsteps in the winter
snow, expelled by the housewife from her worts and leeks, laid
for with a staff or a hare-pipe, and finally doomed to be caught
by the greyhounds and hung up on a pin to be eaten, while the
whelps play with his skin. Only a gentleman gives him a chance,
and that for dread of losing his name:

> For an acre's breadth, he will me see,
>   Or he will let his houndes run.
> Of all the men that be alive,
>   I am most beholden to Gentlemen!

There is a more normal attitude to sport in *The Hunttyng of the
Hare*, if indeed that is not a fake, of which I am not quite
certain. Here the sympathy is all on the side of the dogs,
although the story ends with a battle between their masters.
There are other humorous narratives. The best is *The Turnament
of Totenham*, in riotous *rime couée* stanzas, with much alliteration
and anapaestic rhythm:

> Of all these kene conqueroures to carpe is oure kynde,

says the writer, and justifies his claim by a vigorous burlesque

---

[1] wanhope, 'hopelessness'.

of the usages of chivalry, in which flails take the place of spears.

> Ther was clenkyng of cart sadils and clateryng of cannes:
> Of fel[1] frekis[2] in the feeld brokyn were thaire fannes:[3]
> Of sum were the hedis brokyn, of sum the brayne pannes,
> And euel were they be sene er they went thannes:
> > With swippyng[4] of swipylles.[5]
> > The laddis were so wery forfo3t,[6]
> > That thai my3t fy3t no more on loft,
> > But creppid aboute in the crofte,
> > As thei were crokid crypils.

In one text the fray is followed by a feast in similar vein. *The False Fox* is a beast fable, on a theme used also in Chaucer's *Nun's Priest's Tale*, with a refrain:

> With how, fox, how! with hey, fox, hey!
> Come no more unto oure house to bere our gese aweye!

In *How the Plowman Lerned his Pater Noster* the humour becomes anticlerical. There were scoffers even in the fifteenth century. At the age of forty, the ploughman, although he believes in Christ, does not know the Lord's Prayer. The priest bids him learn it. He goes for shrift, and will give money to be shown how to reach heaven. The priest says that he will send him forty poor men for alms, and double what he gives them. They prove to be *Pater*, *Noster*, and other words of the prayer. All get corn, and the ploughman has none left. He goes to the priest for payment, and the priest only says that he has now paid Christ his rent. The ploughman appeals to a law-court official, but without success. He will never trust a priest again. But he lives and dies peaceably. Sometimes the stories take a ribald turn, as in the *fabliaux* dear to northern France. Of such is *Sir Corneus* or *The Cokwolds Daunce*, which burlesques Arthurian romance. The author calls it a 'bowrd'.[7] Arthur loved cuckolds and had a horn, which spilt if one of them drank from it. They were brought to his table and feasted. He had a visit from the Duke of Gloucester, to whom their case was explained. The horn was brought in. He would not drink before the king, who spilt his draught. The cuckolds laughed; the queen was shamed. They all danced in the 'cokwold rowte'. Arthur lived

---

[1] fel (fele), 'many'.   [2] frekis, 'warriors'.   [3] fannes, 'banners'.
[4] swippyng (swiping), 'swinging'.   [5] swipylles, 'flail-ends'.
[6] forfo3t, 'exhausted with fighting'.   [7] bowrd (bourd), 'jest'.

at Skarlyon with his cuckolds, and had game and glee. Sir Corneus, who served at the king's dais, made this 'gest', and called it after his own name.

> God gyff us grace that we may go
> To heuyn. Amen. Amen.

Even coarser are *A Mery Geste of the Frere and the Boye* and *The Tale of the Basyn.*

A rather widespread type of narrative has a theme which may be called that of the King and the Subject. Oddly enough, one of its favourite incidents is found as early as the twelfth century in the *Speculum Ecclesiae* of Giraldus Cambrensis. Here Henry II dines with an abbot, to whom he is unknown, and they exchange over their cups the toasts *pril* and *wril*, which stand for *wesheil* and *drincheil*. Later the abbot comes to court, and the king astonishes him with *Abbas pater, dico tibi pril*. So it is in the fifteenth-century *King Edward and the Shepherd*. Riding on a May morning near Windsor, the king meets a shepherd, Adam by name, who complains that he has been ill-treated by the royal servants. The king professes to be the son of a Welsh knight. He calls himself Joly Robyn, and says that he has friends at court who will remedy the shepherd's wrongs. The shepherd offers him hospitality. They eat and drink, with further toast-words, as the cup passes, of *passilodion*, which is *wesseyle*, and *berafrynde*. Pressed for venison, the shepherd reluctantly produces some from his cellar. It has, of course, been poached. The king returns home, and Adam boasts to his wife of the powerful friend he has made. Presently he goes to court, expecting to find that his wrongs have been righted. He is graciously received. The king still conceals his identity, but promises him 'gode bourd' and repeats his *passilodion* and *berafrynde*. At last he reveals himself. The shepherd falls on his knees and asks mercy. Here the poem ends rather abruptly, but no doubt all is well. A very similar story, here called a *romans*, is *King Edward and the Hermit*. This, which is located in Sherwood Forest, again turns on the illegitimate venison, produced for the entertainment of a stranger, and has similar toasts, with the words *ffusty bandyes* and *stryke pantere*. Unfortunately the denouement is missing. Similar once more is *John the Reeve*, also a 'bourde', which only comes to us in the seventeenth-century Percy MS., but was known to William Dunbar and

Gavin Douglas in Scotland by the beginning of the sixteenth. Here we get a more complete story. The reeve had reproved his guest over the venison, and when they fell to rough games afterwards,

> Iohn hitt the king ouer the shinnes
> With a payre of new clowted shoes.

Nevertheless, when he gets to court, he is knighted and given a grant of his manor place, with a hundred pounds and a tun of red wine a year to boot, and keeps open house for guests thereafter. Scotland had its own version of this story in *Rauf Coilȝear*, which uses a thirteen-line stanza, with a good deal of alliteration. One other English example, *The King and the Barker*, omits the venison. The king's identity is revealed by the arrival of a noble, who kneels to him. He demands a horse-race, at the end of which the barker is rewarded with a hundred shillings. Similar stories remained popular in England to nearly the end of the eighteenth century. They are mostly found in broadsheets, but about 1599 *The King and the Barker* was turned to dramatic use in the play of *Edward IV*, which may be by Thomas Heywood. The fifteenth-century examples are minstrelsy, as the kind of *incipit* employed makes clear, but minstrelsy of a special type, drawing its material not from chivalric romance but from the life of a lower social grade than that of the courtier. We may perhaps call it yeoman minstrelsy. It must be remembered that at this period there was a great increase in the number of well-to-do yeomen, through the tendency of the bigger landowners to cease the direct cultivation of their demesne fields and to hand them over to leaseholders.

The Sherwood of *The King and the Hermit* is a link between the King and Subject tales and those on the popular hero Robin Hood. Much learned ink has been spent in vain on his origin. There is no reason to identify him with Woden, or the Scandinavian deity Hödr, or the Teutonic elf Hodeken, or to suppose his name a corruption of 'o'th' wood'. Nor are the many Robin Hood's Hills, Wells, Stones, Oaks, or Butts, which are found in all parts of the country, in any way relevant. They are a reflex of his legend, and occur as far from his haunts as Gloucestershire and Somerset. He had no more to do with them than the Devil, to whom similar natural features are also ascribed. 'Hood', in English, can mean nothing but a head-covering. Both Hood and Robinhood are used as surnames

from an early date. A Hód's Oak in an Anglo-Saxon Worcestershire charter may have been a boundary mark of a landowner
so named. A Gilbert Robynhod is found at Fletching, Sussex,
in 1296, a Robert Hode of Newton in Wakefield court rolls from
1316 to 1335, a Robyn Hode as a *vadlet* at court in 1324, a
Robert Robynhoud at Harting, Sussex, in 1332. It seems to
me very likely that the story of the outlaw took its start from a
Robin Hood who in 1354 was in prison awaiting his trial for
trespass of vert and venison in the forest of Rockingham in
Northants. If so, it was not long before, in Thucydidean
phrase, he had won his way to the mythical. By about 1377
the Sloth of *Piers Plowman* claimed to know 'rymes of Robyn
hood, and Randolf erle of Chestre'. About 1401 the *Reply of
Friar Daw Topias to Jack Upland* has the lines

> On old Englis it is said,
> Unkissid is unknowun,
> And many men speken of Robyn Hood,
> And shotte never in his bowe.

This is repeated, almost verbatim, in George Ripley's *Compound
of Alchemie* (1471), and was evidently a proverbial expression.
Chaucer (*Tr. and Cr.* ii. 859) has

> Swich manere folk, I gesse,
> Defamen love, as nothing of him knowe,
> Thei speken, but thei benten nevere his bowe!

About 1410 the author of *Dives and Pauper* reproves those who
'gon levir to heryn a tale or a song of robyn hode or of sum
rubaudry than to heryn messe or matynes'. In 1439 petition
was made to Parliament for the arrest of one Piers Venables of
Derbyshire, who 'in manere of Insurrection wente into the wodes
in that Contre, like as it hadde be Robyn-hode and his meyne'.
But by this time, or even earlier, the fame of the hero had
got as far as Scotland. Walter Bower (*c.* 1385–1449), in the
Chronica Gentis Scotorum, possibly working here on the notes of
John of Fordun (*ob. c.* 1384), makes him one of the dispossessed
followers of Simon de Montfort (*ob.* 1265), but the account of
him evidently comes mainly from literary sources:

Hoc in tempore de exheredatis et bannitis surrexit et caput erexit
ille famosissimus sicarius Robertus Hode et Litill-Johanne cum
eorum complicibus, de quibus stolidum vulgus hianter in comoediis
et in tragoediis prurienter festum faciunt, et, prae ceteris romanciis,

mimos et bardanos cantitare delectantur. De quo etiam quaedam commendabilia recitantur.

He goes on to describe a story which is much like one of those that survive. So too Andrew Wyntoun, in his *Chronicle of Scotland* (*c.* 1420), says of the year 1283:

> Lytill Ihon and Robyne Hude
> Waythemen[1] were commendyd gude,
> In Yngilwode and Barnysdale
> Thai usyd all this tyme thare travale.

In 1438 we hear of a ship called the *Robyne Hude* at Aberdeen. Both in England and in Scotland the name Robin Hood came to be given to the leader of the spring-time folk revels, and in these he is sometimes accompanied by Little John, and sometimes, in England but not in Scotland, by a Maid Marian. Here we may suspect a conflation with the Robin and Marion of the similar French seasonal *jeux*. Robin is also found in some versions of the English mummers' play and in morris-dances.

Further literary allusions to Robin Hood reach us during the latter part of the fifteenth century. In *How the Plowman Lerned his Pater Noster*,

> Eche had two busshelles of whete that was gode,
> They songe goynge home warde a Gest of Robyn Hode.

In a burlesque tale, from a manuscript in a Scottish library,

> The sow sate on hye benke and harpyd Robyn-Hode.

And again,

> Kene men of combur[2] comen belyve,
> For to mote[3] of mychewhat more than a lytull,
> How Reynall and Robyn-Hod runnen at the gleve.[4]

In another manuscript, preserved at Lambeth,

> He that made this songe full good,
> Came of the northe and of the sothern blode,
> And somewhat kyne to Robyn Hode.

But from this period we have some of the tales themselves. They are written in septenars. The earliest is *Robyn and Gandeleyn*, from that Sloane MS. 2593 which yields so many carols. It is 'a carpyng of a clerk', overheard, and has a refrain 'Robyn lyth in grene wode bowndyn'. If not itself strictly a carol, it was probably meant to be sung at a Christmas feast.

---

[1] Waythemen, 'outlawed hunters'. [2] combur, perhaps for Cumberland.
[3] mote (moot), 'tell'. [4] gleve (glaive), 'lance'.

Robin is shot by a 'lytil boy' called Wrennok of Donne, and
Wrennok in turn by Robin's knave Gandeleyn. Neither of
these recurs in connexion with Robin Hood, but Gandeleyn's
name rather suggests the earlier *Tale of Gamelyn*, which is found
in some Chaucerian manuscripts, and which Chaucer possibly
meant to adapt as a tale for his Knight's Yeoman. Gamelyn,
like Robin, is an outlaw, and fights with a sheriff, who is here
his brother. Four other Robin Hood tales may be ascribed,
one of them with some hesitation, as regards its present form,
to the fifteenth century. They all put Robin, not in Rockingham
forest, but either in that of Sherwood, towards the north of
Nottinghamshire, or in that of Barnsdale, in the West Riding
of Yorkshire, bordering upon Nottinghamshire, by the side of
which runs the great highway of Watling Street, from Doncaster
to York. But in all Nottingham itself figures, with its sheriff, who
is Robin's chief enemy. It must be admitted that this worthy
outlaw often appears to be something of a highwayman, as
well as a deer-stealer. But the atmosphere of the greenwood
predominates, and in it Robin's chief associates are Little John,
Will Scarlet, and Much the Miller's Son. It must be added
that Robin is also a man of exemplary devotion to his religious
duties and of great loyalty. The chaotic condition of England
under Edward IV 'our comly Kynge' is well reflected.

In *Robin Hood and the Monk*, which comes to us from a manuscript of about 1450, unfortunately with some gaps in the
narrative, Robin and John rejoice at May-time. Robin will go
to Nottingham to hear mass. Much advises him to take twelve
yeomen, but he will only take John. On the way they shoot at
marks and quarrel over it. Robin goes on, but John returns to
Sherwood. While Robin kneels in St. Mary's Church, a monk,
whom he had once robbed, spies him, orders the town gates to
be sparred, and warns the sheriff, who takes a band to the
church. Robin slays twelve of them, but his sword breaks on
the sheriff's head. He escapes at first, but is ultimately taken.
The monk is sent with letters to inform the king. On the way
he meets with John and Much, who cut off his head and that
of his page. John himself takes the letters to the king who
rewards them, and bids them return to Nottingham with an
order to the sheriff to send Robin to him:

> Ther was neuer yoman in mery Inglond
> I longut so sore to se.

He asks after the monk, and is told that he died by the way. Back at Nottingham, John tells the sheriff that the monk has been made abbot of Westminster. The sheriff feasts John, and himself gets drunk. John goes to the prison, kills the jailer, and releases Robin. They leap the city wall and get back to Sherwood. John tells Robin that he has done him good for evil, and they are reconciled. The sheriff fears the king's wrath, but the king admires John's fidelity to his master:

> Speke no more of this mater, seid our kyng,
> But John has begyled vs alle.

It is a well-told tale, with humour and even some psychology in it. In *Robin Hood and the Potter*, the manuscript of which may be half a century later, the action begins on Watling Street. Robin bets that he will defeat the Potter, who has already defeated John, and get a wad of money from him. He accuses him of not paying pavage, which is toll. They fight, but the Potter wins. They make friends, and Robin borrows the Potter's clothing and some pots, which he will sell at Nottingham. John tells him to beware of the sheriff. Robin disposes of all the pots but five, which he sends as a gift to the sheriff's wife. She asks him to dinner. There is talk of a shooting match. Robin borrows a bow and wins. He tells the sheriff that he has a bow given him by Robin Hood. The sheriff wants to get at that false outlaw. Robin will guide him. When they arrive at the forest, Robin blows his horn, and his men appear. Robin takes the sheriff's horse and gear, but gives him a white palfrey for his wife. She laughs at her husband's discomfiture. Robin gives the Potter ten pounds. In this story humour has prevailed over manslaughter. In *Robin Hood and Guy of Gisborne* two yeomen have taken Robin's bow, and he goes with John in pursuit of them. They find one but quarrel as to who should question him. John goes off to Barnsdale, where he finds two of his men slain, and the sheriff pursuing Will Scarlet. He shoots, but his bow breaks, and only the sheriff's Will a Trent is slain. John is taken and is to be hanged. Meanwhile Robin meets Sir Guy. They have a shooting match, then reveal their names to each other and fight. Sir Guy is slain. Robin puts on the knight's clothes and blows his horn. The sheriff, with the captured John, arrives. He takes Robin to be Sir Guy. Robin says he has slain the master and will now slay the knave.

Instead, he cuts John's bonds, and gives him Sir Guy's bow. The sheriff takes flight, but John cleaves his heart with an arrow. In *Robin Hood and Friar Tuck* Robin and his men are defeated by a pack of dogs belonging to the friar, but Little John shoots them, and the friar becomes one of the outlaw company. These two tales only come to us from the Percy MS. of about 1650. But in some forms they must have existed by about 1475, since both Guy and Tuck appear in a fragmentary piece, not narrative but dramatic, of that date. This probably came from the papers of the Paston family of Norfolk, and by an odd coincidence the Paston Letters contain one of 16 April 1473, in which Sir John Paston laments the loss of a man Woode, whom he had kept 'thys iii yer to pleye Seynt Jorge and Robyn Hod and the Shryff off Notyngham'. So we build up the past. But whether this was one of the comedies and tragedies on Robin Hood themes known to Walter Bower in Scotland before 1449 we cannot tell.

More elaborate than any of these four ballads is *A Lytell Geste of Robyn Hode*, of which an edition was printed by Wynkyn de Worde, not earlier than 1500. It has, however, been suggested that some of the linguistic forms in the text are survivals from Middle English, and may indicate a date for the piece, or for ballads from which it was compiled, even earlier than 1400. A second fragmentary edition comes from the Antwerp press of Jan Van Doesborch (*c.* 1505–30), a third from that of William Copland (*c.* 1560), and a fourth, undated, from that of Edward White. There are some fragments of other editions. Both Copland and White append to the *Geste* a 'playe of Robyn Hoode, verye proper to be played in Maye games', which brings in both Friar Tuck and the Potter. It is distinct from that of 1475, and perhaps also from 'a pastoral pleasant comedie of Robin Hood and Little John &c.', entered on the *Stationers' Register* by White in 1594. The *Geste* may be that referred to in *How the Plowman Lerned his Pater Noster*, and also in the piece with a Reynall in it, noted above from an Edinburgh manuscript. It is very long, running to 456 stanzas, which are divided into eight Fyttes.[1] The action is not very close-knit, and possibly the themes of more than one earlier poem have been conflated. But that of the feud between Robin and the Sheriff of Nottingham still dominates. In the first fytte Robin is in Barnsdale,

---

[1] Fyttes (Fits), 'divisions of a poem'.

with his gang. He will not dine until he has a guest. He is pious. No harm shall be done to husbandmen, but bishops, archbishops, and the sheriff are enemies. The yeomen are sent to Watling Street to look for a guest. They take a knight, whom we learn later to be Sir Richard at the Lee. Robin feasts him. He must pay, but has only ten shillings. Once he had £400, but his son slew a Lancashire knight, and the father's lands are pawned to the Abbot of St. Mary's, York, for his release. He will go on a pilgrimage. Robin will not take God, Peter, Paul, or John for a surety, but will Our Lady. He gives the knight £400 and an equipment, and lends him Little John. In the second fytte the Abbot of St. Mary's is expecting the knight. He has bribed the Sheriff of Nottingham and the High Justice of England to support him. The knight comes and produces the £400, but the High Justice will not return the bribe. The knight goes home to Wyersdale in Lancashire and saves £400 to repay Robin. On his way to hand it over he saves a yeoman from murder at a wrestling. In the third fytte John, calling himself Reynolde Grenelefe, wins a shooting match in the presence of the sheriff, who takes him as a servant. John first fights and then conspires with the cook. They rob the sheriff's treasure and take it to Robin. Then they beguile the sheriff himself while hunting. He has to lie for a night in his breech and shirt under a tree, and is only released on promising to do Robin no scathe. In the fourth fytte Robin again will not dine without an 'unketh¹ gest'. His men take the cellarer of St. Mary's, who has the knight's £400 and £400 more. And when the knight comes to pay his debt to Robin, it is already paid, and he gets the second £400 as a gift. In the fifth fytte the sheriff offers an archery prize, which Robin wins. As he returns to the greenwood the sheriff attacks him and wounds John, whom Robin carries to the knight's castle. In the sixth fytte the sheriff besieges the castle, and goes to the king for justice. The king will come to Nottingham and take Robin. Meanwhile the sheriff catches the knight. His wife appeals to Robin, who goes to Nottingham, cuts off the sheriff's head, frees the knight, and takes him to the greenwood, until he can get grace of 'Edwarde, our comly kynge'. In the seventh fytte the king comes to the north, takes the knight's land, and finds Plumpton Park denuded of deer. He wants vengeance on Robin, and on

¹ unketh (uncouth), 'unknown'.

the advice of a forester disguises himself as an abbot and goes
to the greenwood in search of him. Robin captures him,
relieves him of £20, and gives £20 back. The king shows him
a royal summons to come to Nottingham. They feast on
venison. This is getting very like the King and Subject motive.
At a shooting match Robin misses the mark, and calls on the
king to buffet him. Then he and the knight recognize the king,
and kneel to him. The king bids Robin to come to court. He
will, but unless he likes it, will soon be at the deer again. In the
eighth fytte the king and Robin go to Nottingham together.
Robin's money spent, he gets leave to go back to Barnsdale.
Here he lives twenty-two years, and is at last beguiled to his
death by a kinswoman, the prioress of Kyrkesly Abbey, who for
love of Sir Richard of Doncaster kills him, on pretence of letting
his blood. Of this there was probably a fuller story, but the
earliest version of it which has come down to us is again from
the seventeenth-century Percy MS.

All these fifteenth-century tales of Robin Hood, except that
of *Robyn and Gandeleyn*, must again, like those of the King and
Subject, be classed as yeoman minstrelsy. In *The Monk, The
Potter, Guy of Gisborne*, and *Friar Tuck*, but not in *The Geste*, there
is a seasonal *incipit*, of greenwood leaves in summer. *The Monk*
then plunges *in medias res*. But at the conclusion comes:

> Thus endys the talkyng of the munke
> And Robyn Hode i-wysse;
> God, that is euer a crowned kyng,
> Bryng vs all to his blisse!

In all the others the minstrel at once declares himself. Thus, in
The Potter, he claims audience:

> Herkens, god yemen,
> Comley, corteys, and god.

Later, he interupts his discourse:

> Her es more, and affter ys to saye,
> The best ys beheynde.

And again, at a transition:

> Now speke we of Roben Hode,
> And of the pottyr ondyr the grene bowhe.

And winds up with:

> God haffe mersey on Roben Hodys solle,
> And saffe all god yemanrey!

In *Guy of Gisborne* he passes quickly from a seasonal *incipit*, with a mention of a song-bird's note, to an introduction of his narrative theme.

> And it is by two wight[1] yeomen,
> By deare God, that I meane.

And presently comes:

> Let vs leaue talking of Little Iohn.

So, too, the *Geste* begins:

> Lythe[2] and listin, gentilmen,
> That be of frebore blode;
> I shall you tel of a gode yeman,
> His name was Robyn Hode.

There is courtesy here, for some at least of the audience must have been of villein descent. Before the third fytte the minstrel repeats himself:

> Lyth and lystyn, gentilmen,
> All that nowe be here.

In the middle of the fourth is again a transition:

> ʹNow lete we that monke be styll,
> And speke we of that knyght.

Before the fifth comes again:

> Lyth and lysten, gentil men,
> And herken what I shall say.

And before the sixth:

> Lyth and lysten, gentylmen,
> And herkyn to your songe.

'Songe' must be taken in a loose sense here, for all the other passages suggest that what we have to do with is not sung but recited minstrelsy, a talking, a speaking. Nor is there any hint of a musical accompaniment. It may be added that all the narratives have much of what may be called a minstrel's padding, to eke out stanzas, where rhyme does not come easily, or perhaps merely to emphasize his good faith, with such recurrent phrases as 'For sooth as I the say', or 'Sertenly withouten layn'.[3]

No fifteenth-century narrative poem is ever called in contemporary documents a ballad. Much confusion has been introduced into the history of literature by a loose use of the term. It comes from the French *ballette*, which was a develop-

---

[1] wight, 'stout'.   [2] Lythe, 'Hark'.   [3] layn, 'disguise'.

ment of the earlier danced *chanson de carole*. The linguistic derivation may be from the late Latin *ballare*, to dance. The *ballette* was always lyrical and was limited to three stanzas with a *refrain*. Chaucer followed the model in his *balades*, but often added a fourth stanza, called an 'envoy'. In *The Legend of Good Women* he remembers the dance:

> And after that they wenten in compas,
> Daunsynge aboute this flour an esy pas,
> And songen, as it were in carole-wyse,
> This balade, which that I shal yow devyse.

But other *balades*, by or attributed to him, are of a more general lyrical character, love-poems or complaints or the like. Later poets, such as Lydgate and the Scottish Dunbar, or their scribes, use the term less precisely, often disregarding the limitation to three stanzas, although sometimes appending an envoy. Dunbar, in *The Goldin Terge*, is again reminiscent of the dance.

> On herp and lute full mirrely thay playit,
>   And sang ballattis with michty nottis cleir;
> Ladeis to danss full sobirly assayit.

The *balade* belongs primarily to courtly poetry, although there are more popular examples in a satire on the siege of Calais (1436), which is described as 'yn baladdys', and in a 'balat', which was a prayer, set on the gates of Canterbury in 1460. But to whomsoever addressed, the type remained, throughout the fifteenth century, a lyrical one, in the widest sense, and not a narrative one.

The term underwent a change of sense, with the development of cheap printing in the sixteenth century. Enterprising publishers produced poems of all sorts on folio sheets, using one side of the sheet only, but sometimes juxtaposing two quarto pages of type on it, so that it could be conveniently folded, and generally adding a rough woodcut illustration and an indication of a known tune, to which the poems could be sung. These were sold in shops or handed over to itinerant minstrels, who sang them about the streets and market-places, carrying bundles of copies to dispose of to any purchaser who might be attracted. They are now generally known from their *format* as broadsheets or broadsides. The poems themselves are often described in their titles, obviously by derivation from the fifteenth-century usage, as Ballads. The term is not limited to narrative poems, nor is it invariably used of them. A narrative may be a Song, a Ditty, a Complaint, a Lamentation, a History, a Discourse. The earliest broadside preserved is said to be of 1513. In 1520 John Dorne was selling them in Oxford at a penny each. One of these was on 'Roben Hod'. Many are recorded in the *Stationers' Register*, which begins in 1557, but many, which survive, are not. Probably they were too cheap to be worth paying a registration fee for. Of the ballad-singers we hear a good deal. One has been claimed for the fifteenth century, on the strength of an epitaph:

> Here lyeth under this marbyll ston
> Riche Alane, the ballid man;
> Whether he be safe or noght
> I reche never, for he ne roght.[1]

But 'ballid' at that date can hardly mean anything but 'bald'. From about 1559 we have a manuscript of one Richard Sheale, who lived at Tamworth, and tells us that his business was to 'sing and talk'. It contains a copy of *Chevy Chase*, of which he is not likely to have been the author. Henry Chettle, in his *Kinde-Harts Dreame* (1592), tells us of tradesmen who would send out their former apprentices with a dozen groats-worth of ballads, and of one Barnes in a 'ballad-shambles' or booth, with two sons, who sang 'one in a squeaking treble, the other in an ale-blowne base'. Thomas Nashe abounds in contemptuous references to ballads. Other writers were more sympathetic. Sir Philip Sidney, in his *Apologie for Poetrie* (1595), makes what he calls a confession of his own barbarousness:

I neuer heard the olde song of *Percy* and *Duglas*, that I found not my heart mooued more then with a Trumpet: and yet is it sung but by some blinde Crouder, with no rougher voyce, then rude stile.

So, too, Sir William Cornwallis, in his *Essayes* (1600), admits:

I have not beene ashamed to aduenture mine eares with a balladsinger, and they haue come home loaden to my liking, doubly satisfied, with profit, & with recreation. The profit, to see earthlings satisfied with such course stuffe, to heare vice rebuked, and to see the power of Vertue that pierceth the head of such a base Historian, and vile Auditory. The recreation to see how thoroughly the standers by are affected, what strange gestures come from them, what strained stuffe from their Poet, what shift they make to stand to heare, what extremities he is driuen to for Rime, how they

[1] roght (recked), 'cared'.

aduenture their purses, he his wits, how well both their paines are recompenced, they with a filthy noise, hee with a base reward.

Naturally, the ballad-singers provided mirth for the stage-players, who were in some sort their rivals. In *1 Henry IV* Falstaff, tricked by Prince Hal and his companions, threatens to have ballads made on them, and sung to filthy tunes. In *2 Henry IV*, after claiming the capture of a knight, he will have it put in a ballad, 'with mine own picture on the top on't, Colville kissing my foot'. In *Antony and Cleopatra* the queen anticipates her presence with Iras at Caesar's triumph in Rome, when 'scald rhymers' will 'ballad us out o' tune'. A full-blown portrait is of course the Autolycus of *The Winter's Tale*, who was pedlar as well as ballad-hawker. In Ben Jonson's *Bartholo-mew Fair* Nightingale is a 'sweet singer of new ballads allurant', and Squire Cokes recalls 'the ballads over the nursery chimney at home of my owne pasting up'. Ballads became the literature of the people. Robert Laneham, describing Queen Elizabeth's visit to Kenilworth in 1575, tells of Captain Cox, who had a bunch of more than a hundred of them 'fair wrapt up in parchment, and bound with a whipcord'. In the seventeenth century Richard Corbet, already by 1617 a Doctor of Divinity, was at the Cross in Abingdon on a market-day:

The ballad singer complaynd, he had no custome, he could not putt-off his ballades. The jolly Doctor putts-off his gowne, and putts-on the ballad singer's leathern jacket, and being a handsome man, and had a rare full voice, he presently vended a great many, and had a great audience.

John Aubrey, born in 1626, tells us that his nurse 'had the history of the Conquest down to Carl I in ballad'. Piscator, in Izaak Walton's *Compleat Angler* (1653) says:

I'l now lead you to an honest Ale-house, where we shall find a cleanly room, Lavender in the windowes, and twenty Ballads stuck about the wall.

In the same year Dorothy Osborne, at Chicksands in Bedfordshire, would walk in a common 'where a great many young wenches keep Sheep and Cows, and sitt in the shade singing of Ballads'.

Most writers of broadside ballads remain anonymous, but amongst them were William Elderton (1559–84) and Thomas Deloney (1543?–1600?), whom Thomas Nashe calls 'the ballet-ing silk-weaver'; and with Deloney's *Garland of Good Will* begins a practice of occasionally publishing ballads in small pamphlets, which continued to be called Garlands. Deloney also wrote prose narratives, and in his *Pleasant History of John Winchcomb* (1597) introduced a ballad of *The Fair Flower of Northumberland*, of which he says, 'the maidens in dulcet manner chanted out this song, two of them singing the ditty, and all the rest bearing the burden'. Scotland, like England, had its broadsides. Robert Lekpreuik was printing them by 1561, Henry Charteris by 1580, and Andro Hart by 1610. They were doubtless looked on with disfavour by the ecclesiastical authorities, and in 1567 an alternative was offered in *Ane Compendious Buik of Godlie Psalms and Spiritual Sangs, collectit furthe of sundrie partis of the Scripture, with vtheris Ballates changeit out of prophane Sangis in Godlie sangis, for avoyding of sin and harlotrie.* The authorship is attributed to James, John, and Robert Wedderburn. The contents, however, are lyrical, not narrative. The ballad-singers must, I think, be taken to have inherited the tradition of medieval popular minstrelsy. How far the minstrels, who continued to be borne on the establishments of courts and castles, still occupied themselves with narrative poetry must remain obscure. Probably they were largely instrumentalists. But they occasionally, in the sixteenth century at least, travelled abroad. Elizabethan *Statutes of Vagabonds* provide for the whipping of wandering minstrels who have not a licence from Justices of the Peace, but make an exception for those belonging to barons or higher nobles, and minstrels on the road were likely enough to vary the entertainments they gave. Henry Algernon Percy, Earl of Northumberland, had on his Household Book about 1512 a Tabarette, a Luyte, and a Rebecc, as well as six Trumpeters, and it would be hard if they could not among them put together a ballad on some exploit of the Percy family. The great families of the north were proud of their past. From the Stanleys of Lancashire we have two historical pieces, one possibly from the fifteenth century, which look like the work of a retainer. And about 1562 Thomas Stanley, Bishop of Sodor and Man, composed a metrical chronicle of his house.

Broadsides were ephemeral things. Many have, no doubt, been lost, but many have been preserved through antiquarian curiosity. The earliest collection, at Shirburn Castle in Oxfordshire, comes from about 1600 to 1616. It is in manuscript, but variations in the characters used by the writer show clearly

that it was copied from black-letter prints. In 1641 a gentleman
from Worcestershire said to John Prideaux, at his consecration
as Bishop of Oxford:

> Lend me what ballads you have, and I will let *you* see what I have:
> I know you to love all such things.

Doubtless broadsides contributed to the Percy Folio MS., which
dates from about 1650. It owes its name to Thomas Percy,
who became Bishop of Dromore. He was not related to the
Percies of Northumberland, although he liked to think he was.
Born at Bridgnorth, Shropshire, in 1729, he found the manuscript during his youth in the house of Humphrey Pitt of Shifnal,
also in Shropshire. It had been mutilated by maidservants who
had torn out leaves for the lighting of fires. A hundred and
ninety-one poems, narrative and lyrical, survive. Of the narratives a considerable number are not found elsewhere, but of a
few traditional versions have been gathered in Scotland. All
appear to have been edited by a single hand, writing in English
dialect, and with a constant trick of introducing lines of dialogue
with 'Sayes' or 'Quoth' or equivalent openings. The compiler
seems to have been interested in the Border and in the Percy
and Stanley families. The political sympathies of the narratives
are English. Extensive collections of broadsides were made by
John Selden (1584–1654), Anthony Wood (1632–95), Samuel
Pepys (1633–1703), John Bagford (1650–1716), and Robert
Harley, Earl of Oxford (1661–1724).

It has often been forgotten that, both in England and in
Scotland, the printing of broadsides continued into the nineteenth century. 'Stall ballads' they are sometimes called. An
interesting light is thrown upon this by Mr. Alfred Williams
in his *Folk-Songs of the Upper Thames* (1923). Out of the six
hundred songs which he has obtained, he does not believe that
over ten or twelve were composed in the Thames valley itself.
A few, on local events, were printed at Wotton-under-Edge,
and some others at Cirencester and Highworth. And he adds:

> At the same time, I am certain the pieces were not composed in
> the locality. The ballads were probably out of print and unprotected
> by copyright. I have also sheets that were printed at Bristol,
> Newport, Birmingham, Winchester, and London. The majority of
> the songs and ballads, in my opinion, were written in London and
> other large towns and cities. There appears to have been a school
> of such ballad-writers, very well trained to their work, and admirably

informed as to the best means of captivating the ear of the public. No doubt the work was remunerative. We know that enormous quantities of the sheets were sold up and down the countryside; hundreds, if not thousands, were commonly disposed of at a single fair-time.

Later he says:

The songs were mainly obtained at the fairs. These were attended by the ballad-singers, who stood in the market-place and sang the new tunes and pieces, and at the same time sold the broadsides at a penny each. The most famous ballad-singers of the Thames Valley, in recent times, were a man and woman, who travelled together, and each of whom had but one eye. They sang at all the local fairs, and the man sold the sheets, frequently wetting his thumb with his lips to detach a sheet from the bundle and hand it to a customer in the midst of the singing.

Mr. Williams began his collecting in 1914, and himself knew a man who, being unable to read, memorized a long song which he had only heard once, at Highworth Fair. Broadsides continued to be printed in London by James Catnach and others up to the end of the nineteenth century. A little earlier they were still coming also from northern towns.

Joseph Addison, writing on *Chevy Chase* and *The Two Children in the Wood* in *The Spectator* for 1711, described ballads as 'the darling Songs of the common People'; and during the eighteenth century it gradually became recognized that, especially in Scotland, many verse narratives existed in traditional form and could be recovered, although often only in a fragmentary condition, from the lips of peasants. Broadsides were probably still the main source for a three-volume *Collection of Old Ballads* (1723–5), the editorship of which is generally ascribed to Ambrose Philips. This was followed in Edinburgh by Allan Ramsay's pioneer contribution in the three instalments of his *The Tea-Table Miscellany* (1724–7), which contained both Scottish and English poems. But it was not until 1757 that Thomas Percy began to contemplate making use of his Folio MS. A long correspondence followed with the poet William Shenstone, who in the course of it suggested the desirability of using the term Ballad only for a poem 'which describes or implies some action', and calling one 'which contains only an expression of sentiment' a Song. This convenient dichotomy met with acceptance. Finally the Folio MS. became the chief source for Percy's *Reliques of Ancient English Poetry* (1765), although much was added both from broadsides and from traditional texts sent to the editor by various correspondents, including Scottish ones from Sir David Dalrymple, Lord Hailes. There was some fragmentary material to deal with, and Percy's manner of handling it was not in accordance with modern canons of editorial fidelity to the sources. As he himself put it later, he thought that *lacunae* could be filled up 'in the manner that old broken fragments of antique statues have been repaired and compleated by modern masters'. The full extent of this sophistication was not apparent until the Folio MS. was itself reproduced by J. W. Hales and F. J. Furnivall in 1867–8. But Percy's methods did not escape the lynx eye of the learned but combative Joseph Ritson, who castigated them in the preface to his own *Select Collection of English Songs* (1783). In *The Gentleman's Magazine* for 1784 Ritson meted out similar treatment to John Pinkerton, whom he frankly, and not without some justification, denounced as a forger in his *Scottish Tragic Ballads* (1781) and *Select Scottish Ballads* (1783). It must be admitted that most of the collectors of Scottish ballads, who were inspired by the *Reliques*, were lovers of poetry rather than exact scholars, and did not refrain from completing for themselves pieces which had reached them from singing or recitation in a fragmentary condition. Sometimes, too, they were tempted to 'correct' what they found unintelligible. This is true even of Sir Walter Scott, in his *Minstrelsy of the Scottish Border* (1802–3), the fruit of ten years' hunting among the farms of Liddesdale. The worst offenders have generally been taken to be Peter Buchan and the blind stroller James Rankin, who supplied him with the material. Mr. Keith has recently done something to rehabilitate the reputation of these worthies. It is fortunate, however, that some of the texts, upon which Scott and others worked, have also been preserved in the form in which they were originally taken down.

An edition of Percy's *Reliques* was printed at Göttingen in 1767, and led to much German speculation on the distinction already made in the sixteenth century by Montaigne, between popular poetry and the poetry of art. J. G. von Herder translated a number of the ballads, and apparently introduced the term *Volkslied*. In an essay on Homer he wrote of an age when in the mouths of the people heroic traditions 'of themselves took on poetic form', but was careful to make it clear that by 'the people' he meant the whole of a race and not merely 'the rabble of the streets'. He was followed by A. W. Schlegel, who wrote less cautiously of ballads, 'Deren Dichter gewissermassen das Volk im ganzen war'. And in later German speculation the notion of composition by the folk took on a more mystical form. This was largely due to the activity of the brothers Jakob and Wilhelm Grimm, although the phrase *Das Volk dichtet*, in which their theory is often summed up, does not appear to have been actually used by either of them. But Jakob Grimm certainly committed himself to the view that 'every *epos* must compose itself, must make itself, and can be written by no poet'. Later he admitted that the exact process by which the self-composed epic came into existence was not quite clear to him.

Epic poetry is not produced by particular and recognized poets, but rather springs up and spreads a long time among the people themselves, in the mouth of the people—wie man das nun näher fasse.[1]

Wilhelm Grimm applied his brother's theory, rather more cautiously, to the ballad. And now Schlegel, in spite of his own earlier utterance, protested. Every poem, he held, implies a poet. Legend and *epos* and song might well belong to the people as their property; but the making of this verse was never a communal process. Later the controversy was renewed in connexion with the origin of Scandinavian ballads. Ferdinand Wolf insisted that they were made 'von einem dichtenden Subjekt', and not 'von einem nebulosen Dichteraggregat, Volk genannt', but was countered by Svend Grundtvig with

Darum ist das Volks-Individuum als solches, nicht das einfache Menschen-Individuum, als Dichter der Volkspoesie zu betrachten.

And there, for a while, the issue between communalists and individualists may rest.

The outstanding collection of ballads is that by Professor F. J. Child of Harvard. Of this a first edition appeared in 1857–8. It was expanded, after many years of further study, into the magnificent *corpus* of *The English and Scottish Popular Ballads* (1882–98). Here Child gave all the versions known to him of 305 ballads, with their textual variants in different copies and an introductory comment on each ballad. The versions are often very numerous, running to over ten for over thirty

---

[1] 'However we may interpret this phrase more precisely.'

ballads, and to over twenty for *The Twa Sisters*, *Lamkin*, *Sir Hugh*, and *Marie Hamilton*. Child was able to correct many sophistications of the Scottish collectors, by reference to their original manuscripts, several of which are now in the Harvard Library. Some fresh traditional material he obtained from his coadjutor Mr. W. Macmath. Unfortunately Child died in 1896. The last parts of his collection were edited by the competent hand of Professor G. L. Kittredge, but the general introduction, with which Child would doubtless have completed his work, was never written. In 1904 Kittredge wrote one for a shorter collection by himself and Miss H. C. Sargent, of which an English edition appeared in 1905. This was based upon Child's, and included all his 305 ballads, except five, which were apparently thought too 'high-kilted' for the general public. Child himself had approved the plan. Generally only one version is given, sometimes two or three, with short introductory notes. Child's collection was remarkably complete so far as the material available in his day went.

Ballad-collecting did not, however, stop with Child's lifetime. Gavin Greig swept together the last leaves of Scottish tradition in Aberdeenshire. Interesting, too, were the researches of Cecil Sharp and his colleagues of the English Folk-Song Society, which disclosed the survival of much unsuspected balladry in southern and midland England, thus to some extent qualifying the preponderance of northern elements in Child's collection. Further evidence of this was provided by the gatherings of Williams in the upper Thames valley. Perhaps the most valuable of the new discoveries was that of *The Bitter Withy*, first unearthed by Mr. Frank Sidgwick in Herefordshire and later traced in several other English counties. This, with the analogous *Seven Virgins*, made a considerable addition to the small number of religious ballads known to Child. More surprising was the revelation of a tradition of balladry inherited through settlers from this island in northern America. Here a circular of inquiry issued by Child in 1881 had met with little response. But a visit of Sharp and Mrs. Campbell to the remote and almost inaccessible area of the Appalachian mountains, to the southwest of Washington, led to an unexpected result. Here were a number of farming communities whose members had been cut off from the surrounding civilization for a hundred years. They still spoke an old-fashioned English dialect and, although illiterate, had a culture of their own, which consisted largely of ballads, presumably inherited by oral tradition from ancestors who had migrated from England and the Scottish lowlands during the eighteenth and the early nineteenth centuries. They had no broadsides, but children sometimes had written copies, which they still called 'ballets'. Since then, the researches of American scholars have revealed the existence of ballads in many other parts of the United States. They are mainly found in rural districts, sometimes among lumberjacks and cowboys, and even among native Indians and imported negroes. Many are versions, often much perverted, of English and Scottish ballads, but others have more indigenous themes of emigrant and frontier life.

As to Child's own views on the origin and nature of balladry, we have little to go upon beyond the contents of the collection itself and an article contributed to Johnson's *Encyclopaedia* in 1893, which he did not wish to be regarded as 'final'. According to one of his critics, 'he never made clear the standard by which he ruled out or included a given specimen'. No doubt there was some sort of Platonic 'idea' of the ballad in his mind. The 'popular' of his title must, I think, refer to usage rather than to authorship. It excludes narrative poems of courtly or literary types. But it was certainly not meant to exclude minstrel ballads. There is much minstrelsy in Child's volumes. Since his day 'ballad' without any epithet has practically come to mean a 'Child' ballad, or one of similar character more recently discovered. This is a limitation of Shenstone's proposed use of the term, to cover all narrative as distinct from lyrical poems. It is convenient enough. Ballads, in Child's sense, are of course all narratives. Most, although not all, of them are lyrical narratives. But I think it is putting the cart before the horse to treat them as narrative lyrics. The learned W. P. Ker, who wrote much about ballads, does not seem ever to have been quite clear on this point of classification. In a lecture of 1909 to the British Academy he said, 'Ballad is here taken as meaning a lyrical narrative poem', adding that 'all ballads are *lyrical* ballads', and then again:

It is not a narrative poem only; it is a narrative poem lyrical in form, or a lyrical poem with a narrative body in it.

In some later lectures, if he is correctly reported, he seems at one point to have come down on the lyrical side, defining a ballad as 'a song with a story in it', but in the same series to have spoken of the Border ballads in particular as 'giving a story, and lyrical only in the stanza'. So, too, Professor Gummere wrote, 'The ballad, of course, like any real lyric, must be sung, it must have a tune'; and so, too, in his turn, Professor Gerould, almost in Ker's words, 'A ballad is a folk-song that tells a story', and in amplification,

What we have come to call a ballad is always a narrative, is always sung to a rounded melody, and is always learned from the lips of others rather than by reading.

Professor Entwistle is more cautious, if less consistent. For him, too, in one place 'a ballad is not a ballad except when sung', but elsewhere 'in England a recitative style seems to have been customary'. Certainly many ballads, possibly most ballads, have at some time been sung. But not all ballads. One of our earliest English ballads, *Robin Hood and the Monk*, has already been noted above as a 'talking'. And this may well be the best description of other long minstrel ballads, even if they were given in a chanting tone, and perhaps with some musical accompaniment. When Erasmus visited England he wrote of English minstrels, who *recitant* but *studio vitant cantum*. No doubt there were exceptions. Richard Sheale's business was both to sing and talk. The broadside printers provided tunes, although not always very suitable ones. Many ballads from Scottish tradition have come to us through singing, but many others from recitation. Mrs. Brown of Falkland, although she learnt her ballads from the singing of her mother or an old maidservant, herself recited them. Many of Greig's Aberdeen ballads had passed through two generations of recitation. It is not legitimate to take it for granted that a period of singing had always preceded the recitation. The ballads recently recovered from among the English folk are sung, often to melodies in old-fashioned forms, known as the 'modes', which were once used in ecclesiastical music, but have long been abandoned by learned composers. It is clear, however, from the experience of those who have studied the methods and attitudes of folk-singers that in their minds the essential feature of a ballad is the story which is related, and not the melody which accompanies it. On English evidence Cecil Sharp wrote:

It is a well-known fact that the folk-singer attaches far more importance to the words of his song than to its tune; that, while he is conscious of the words that he is singing, he is more or less unconscious of the melody.

And of the Appalachian singer he says much the same:

So far as I have been able to comprehend his mental attitude, I gather that, while singing a ballad, for instance, he is merely relating a story in a peculiarly effective way which he has learned from his elders.

Gavin Greig's experience in Aberdeenshire was contrary to Sharp's on this point. I feel sure, however, that any definition of a ballad ought to be primarily in terms of the narrative which it records, and not in those of the lyrical element which accompanies it.

Ballads, with a few exceptions, are written either in four-stress rhyming couplets or in septenars, which were probably felt rather as quatrains, alternately of four-stress and three-stress lines, of which only the three-stress ones rhyme. The rhythm is normally iambic, but there is often a feminine ending to a line, and often an unstressed syllable lacking before the first stressed one, which gives something of a trochaic effect. Often again a stressed syllable carries with it more than one unstressed one. According to Professor Gerould, whatever the metre, heavy and light stresses follow each other in regular succession. This, he says, is more clearly apparent from ballad tunes than from the verbal texts. The stanza proper, whether a couplet or a quatrain or more elaborate, is often accompanied by a refrain, which is of course itself a lyrical element. It has been calculated that a refrain occurs in at least one version of 106 of Child's 305 ballads, and that of all of his ballad versions about a quarter have a refrain. The proportion in ballads recently recovered from folk-singers is more like a half. Most commonly the refrain consists of two lines alternating with those of a stanza of four-stress couplets. But sometimes each narrative stanza is followed by a stanza or a single line of refrain. There seems no justification for Child's assumption that the four-stress stanza was the earliest type of ballad metre, or for Professor Gummere's corollary that at first ballads 'always had a refrain'. And certainly there is no evidence, beyond Deloney's account of *The Fair Flower of Northumberland*, that ballad refrains were sung by a chorus. Ballad refrains are very different in character from those of carols, which are essential to the themes and are logically linked to the stanzas which introduce them. In ballads, on the other hand, the refrains are often mere nonsense jingles. A score of the Robin Hood ballads from broadsides have, with slight variations, the refrain:

> Down a down a down a down.

Other examples, in ballads of a more traditional type, are:

> With a down derry, derrie, derrie, down, downe.

And:

> Fa la la la, fa la la la ra re.

And:

> Sing trang dil do lee.

And:

> Oh and allelladay, oh and allelladay.

Such refrains rather look as though they might be substitutes for musical accompaniments. An extreme case is in the humorous ballad of *The Whummill Bore*, the first stanza of which runs:

> Seven lang years I hae served the king,
>   Fa fa fa fa lilly,
> And I never got a sight of his daughter but ane.
>   With my glimpy, glimpy, glimpy eedle,
>   Lillum too tee a ta too a tee a ta a tally.

Another is in *The Maid and the Palmer*:

> The maid shee went to the well to washe,
>   Lillumwham, lillumwham!
> The mayd shee went to the well to washe,
>   Whatt then? what then?
> The maid shee went to the well to washe,
>   Dew ffell of her lilly white fleshe.
>   'Grandam boy, grandam boy, heye!
> Leg a derry, leg a merry, mett, mer, whoope, whir!
> Driuance, larumben, grandam boy, heye!

Here the couplet is submerged under the repetition of one of its lines, and the refrain dominates. There are, of course, refrains which help directly to advance the narrative theme of a ballad. Others again introduce a descriptive touch of natural beauty in garden or moorland, which may add poignancy to a tragic happening, or emphasize a more sentimental one. Such are:

> With a hey and a ho and a lillie gay,
> As the primrose spreads so sweetly.

And:

> Hey nien nanny
> And the norlan flowers spring bonny.

And:

> Gentle Jenny cried rosemaree,
> As the dew flies over the mulberry tree.

'Dew' should be 'Dow', the Scottish form of 'Dove', I suppose. Refrains sit lightly to their ballads. Some recur in more than one ballad, and different versions of the same ballad often have different refrains. Two tragic ballads, for example, have the pretty refrain:

> The broom blows bonnie and so is it fair,
> And we'll never gang up to the broom nae mair.

Sometimes it is difficult to trace any relevancy of a refrain to the ballad which it accompanies. Thus a version of *The Twa Sisters* begins:

> There was twa sisters in a bowr,
>   Edinburgh, Edinburgh,
> There was twa sisters in a bowr,
>   Stirling for ay,
> There was twa sisters in a bowr,
> There came a knight to be their wooer.
> Bonny Saint Johnston stands upon Tay.

But the subject of the ballad is the drowning of a girl in a mill-dam, which another version of the ballad puts in Binnorie, and although I do not know where Binnorie is, it is hard to see how Edinburgh and Stirling and St. Johnston on Tay can all have been concerned. I have a strong suspicion that scraps of independent lyrics were often adopted as ballad refrains.

Child's collection is not a homogeneous one, and it will be as well to examine it in detail, both as regards chronology and as regards the various types of balladry included. Chronologically it begins with *Judas*, which comes from a manuscript of the end of the thirteenth century. It will be as well to give the text. I follow Mr. Kenneth Sisam's transliteration, which seems more precise than Child's. The piece appears to be written in a mixture of septenar and Alexandrine lines, of which there are other thirteenth-century examples.

> Hit wes upon a Scere[1] Thorsday that vre Louerd aros;
> Ful milde were the wordes He spec to Iudas:

---

[1] Scere, 'Holy'.

'Iudas, thou most to Iurselem, oure mete for to bugge;[1]
Thritti platen[2] of seluer thou bere upo thi rugge.[3]

Thou comest fer i the brode stret, fer i the brode strete;
Summe of thine cunesmen[4] ther thou meist imete.'

Imette wid is soster, the swikele[5] wimon:
'Iudas, thou were wrthe[6] me stende[6a] the wid ston,

Iudas, thou were wrthe me stende the wid ston,
For the false prophete that tou bileuest upon.'

'Be stille, leue soster, thin herte the tobreke![7]
Wiste min Louerd Crist, ful wel He wolde be wreke.'[8]

'Iudas, go thou on the roc, heie upon the ston,
Lei thin heued i my barm,[9] slep thou the anon.'

Sone so Iudas of slepe was awake,
Thritti platen of seluer from hym weren itake.

He drou[10] hymselve bi the top,[11] that al it lauede[12] a blode;
The Iewes out of Iurselem awenden[13] he were wode.[14]

Foret[15] hym com the riche Ieu that heiste[16] Pilatus:
'Wolte sulle thi Louerd, that hette[17] Iesus?'

'I nul sulle my Louerd for nones cunnes eiste,[18]
Bote hit be for the thritti platen that He me bitaiste.'[19]

'Wolte sulle thi Lord Crist for enes cunnes golde?'
'Nay, bote hit be for the platen that He habben wolde.'

In him com ur Lord gon, as is postles seten at mete:
'Wou sitte ye, postles, ant wi nule ye ete?

Wou sitte ye, postles, ant wi nule ye ete?
Ic am iboust ant isold today for oure mete.'

Vp stod him Iudas: 'Lord am I that?
I nas neuer o the stude[20] ther me The euel spec.'

Vp him stod Peter, ant spec wid al is miste:
'Thau Pilatus him come wid ten hundred cnistes,[21]

Thau Pilatus him come wid ten hundred cnistes,
Yet ic wolde, Louerd, for Thi loue fiste.'[22]

'Stille thou be, Peter! Wel I the icnowe;
Thou wolt fursake me thrien[23] ar the coc him crowe.'

---

[1] bugge (buy).   [2] platen, 'pieces'.   [3] rugge, 'back'.   [4] cunesmen (kinsmen).
[5] swikele, 'treacherous'.   [6] wrthe (worthy).   [6a] me stende, 'men stoned'.
[7] the tobreke, 'give thee remorse'.   [8] wreke, 'revenged'.   [9] barm, 'lap'.
[10] drou, 'tore'.   [11] top, 'hair of head'.   [12] lauede, 'poured'.
[13] awenden, 'thought'.   [14] wode, 'mad'.   [15] Foret, 'Before'.
[16] heiste, 'was named'.   [17] hette, 'is named'.
[18] nones cunnes eiste, 'no kind of goods'.   [19] bitaiste, 'entrusted'.
[20] stude, 'place'.   [21] cnistes (knights).   [22] fiste (fight).   [23] thrien (thrice).

Whether the poem is quite complete I do not feel sure; it seems
to end very abruptly. It is true that, as Mr. Sisam says, it is
'the only example before 1400 of the swift and dramatic movement, the sudden transitions, and the restrained expression,
characteristic of the ballad style'. Nevertheless I do not think
that it can be properly regarded as a ballad, certainly not as
a 'popular' ballad. It is one of a group of religious poems, of
which another on Twelfth Day has also been claimed, although
not by Child, as a popular ballad, or possibly only a 'literary
imitation' of one. Both poems are in the same hand, but of
that on Twelfth Day the manuscript also contains a rough draft
in red chalk, partly overwritten, which at least suggests that the
scribe was also the author of both. I feel sure that they are of
purely ecclesiastical origin. It is obviously *a priori* improbable
that a literary type should emerge in the thirteenth century
and not be traceable again before the fifteenth. Child includes
in his collection *Robyn and Gandeleyn* and *St. Stephen and Herod*,
from Sloane MS. 2593, which may be of the first half of the
latter century. I have already expressed a doubt whether *Robyn
and Gandeleyn*, which has a refrain, may not have been intended
for use as a carol. *St. Stephen and Herod* is not in carol form, but
I think that it was meant for singing in a hall at Christmas.
It has references to the 'boris hed' of the feasts, and to the
saint's 'euyn on Crystes owyn day' and even a bit of Latin in
'The capoun crew *Cristus natus est*'. These things are hardly
in the ballad manner. If I am right, then, the first really
'popular' ballads in Child's collection of which we can be sure
are *Robin Hood and the Monk* and *Riddles Wisely Expounded*, both
of which come from manuscripts of about 1450. Then comes
Robin Hood and the Potter from one of about 1500. It is possible
that *A Lytell Geste of Robyn Hode* may be earlier, in view of its
apparent retention of some Middle English forms. And some
version of *Robin Hood and Guy of Gisborne*, which we have only
from the Percy MS., must have existed at least as early as 1473.
Child does not include any of the King and Subject narratives,
except a *King Edward and a Tanner of Tamworth*, which is a late
and much abridged version of *The King and the Barker*.

It is regrettable that recent historians of literature have not
been careful to restrict the meaning of the term 'ballad' to
narrative poems of the 'Child' type, but have used it freely of
the medieval *carmina* and *cantilenae* recorded by William of

Malmesbury and others, about which I have already written something in connexion with carols, and of which we often do not even know whether they were lyrical or narrative in character. Even the lays supposed to lie behind the Anglo-Saxon epics have sometimes been called ballads. Professor Gerould is more careful here in his terminology than is Professor Gummere, even in his latest utterance. But perhaps the writer most open to criticism in this respect is the very latest exponent of balladry, and a very learned one, who takes all Europe for his range, Professor Entwistle. English balladry, he tells us, began either in the mid-twelfth century or later. This is a safe statement enough, if we do not forget the 'or later'. But elsewhere Professor Entwistle appears to do so:

The cardinal document as to the date of our ballads is a piece of external evidence. William of Malmesbury states definitely that a poem about Canute's daughter Gunhild, falsely accused before her husband the Emperor Henry III and unexpectedly delivered, was 'nostris adhuc in triviis cantitata' (c. 1140). Brompton (c. 1350) names her accuser and defender, Roddyngar and Mimicon; Matthew of Westminster gives us Mimecan. There is no doubt that these references are to a poem of traditional nature and of content identical with the ballad of *Sir Aldingar*. The poem was either the ballad itself, or some very similar piece in another style which we gratuitously hypothetize.

And later he adds:

In Saxon England there was a ferment of creation, of which the ballad of *Sir Aldingar* is one fruit.

But surely there could be no more gratuitous hypothesis than an assumption that a poem which, like *Sir Aldingar*, comes to us from the Percy MS. of about 1650 can be identical in style with one known to William of Malmesbury in the twelfth century. This indeed, so far as diction goes, Professor Entwistle in another passage admits. As regards the substance of the narrative, it is true that, as he points out, there is a link of common tradition. William of Malmesbury describes Gunhild's accuser as being *giganteae moliminis*[1] and the Queen's champion as *puerulum*, and in *Sir Aldingar* they are respectively 'as bigge as a ffooder'[2] and 'a little child'. In a Scottish version of the ballad the accuser, here called Rodingham, is 'stark and stoor', but the champion is merely a knight from the north, Sir Hugh le

---

[1] *giganteae moliminis*, 'of gigantic bulk'.   [2] ffooder (fother), 'cartload'.

Blond. In neither is the heroine called Gunhild. In *Sir Aldingar* she is Elinor 'our comly queene', and her husband is Harry, 'our comlye king'. They are of England, not Germany. Surely this does not come from William of Malmesbury. He was dead before any English King Henry married an Elinor. Probably the ballad-writer had in mind Elinor of Aquitaine, the wife of Henry II. But the false accusation of unchastity against a queen is an ancient story. It was told of the Lombard queen Gundiberga as early as the seventh century, and later of St. Cunigund, the wife of the Emperor Henry II. In 1338 the *joculator* Herbertus told it at St. Swithin's, Winchester, of Emma, the wife of Canute, and this version may have been known to Langland, who quotes a line *Dieu vous sauve, Dame Emme*, which looks like a refrain. The theme is widespread in Scandinavia, and here the heroine is generally Gunhild. And in one important version, the Danish *Ravengaard og Memering*, the champion is Mimecan. Professor Entwistle returns to the question of chronology in connexion with some other ballads. 'The age of British balladry', he here says, 'is hard to determine', but 'there must have been heavy losses of medieval material.' But he thinks that a secure date can be given to *The Battle of Harlaw*:

The event occurred in 1411, and it was of immediate local interest to Aberdonians. The ballad is of the sort which arises directly out of the experience it narrates, and we are, in any case, certain that it existed in 1549.

We are, no doubt, but I do not see how this helps the argument that it existed more than a century and a quarter earlier. Moreover, it falsifies history, since it was the Highlanders and not, as it claims, the Lowlanders, who had the better of the day. Of Robin Hood, Professor Entwistle says:

We find that Robin Hood was probably a real personage of the twelfth century (dying in 1198), and that four ballads were stitched into the *Geste* before 1400. The circulation of Robin Hood pieces is attested in 1377; nothing prevents the supposition that they were composed much earlier.

Nothing prevents any supposition, but a supposition is not evidence. It is possible that some Middle English forms may survive in the *Geste*. The pieces of 1377 are only called 'rymes' The date of 1198 only rests on the statement of Martin Parker in his *True Tale of Robin Hood*, printed in 1632, that the date was to be seen 'within these hundreth yeares' on a tombstone at Kirklees Abbey, with an epitaph which described Robin as Earl of Huntingdon. This is not material for literary history. There are in fact eight historical or quasihistorical ballads, on actual or legendary events, belonging or in some cases only ascribed by conjecture to centuries earlier than the fifteenth. The earliest is *Queen Eleanor's Confession*. Here we may again have Henry II (*ob.* 1189) and Eleanor of Aquitaine (*ob.* 1204), although in this case the marital unchastity is represented as actual and not, as in *Sir Aldingar*, merely the subject of a trumped-up charge. The ballad can hardly be of early date, as the confession is heard by two friars, which is not in accordance with Catholic practice.

To the thirteenth century belongs the legendary theme of *Sir Hugh* and also, if a very doubtful conjecture is accepted, that of *Sir Patrick Spens*. From the fourteenth come those of the battles of *Durham Field* (1346) and *Otterburn* (1388), and probably those of *Hugh Spencer's Feats in France* and of the fragmentary *Knight of Liddesdale*. Of most of these Professor Entwistle says, 'Their testimony is not conclusive, but, taken cumulatively, it seems to establish the existence of English ballads in the fourteenth century, while leaving open the question of an earlier date.' We must not, however, forget that there were other channels, besides songs, for the transmission and perversion of historical and legendary themes, through oral tradition, easily long-lived in monasteries, and still more through chronicles in prose and verse. Certainly some of the ballad-writers drew upon chronicles. 'The cronykle will not layne',[1] says the author of *The Battle of Otterburn*, and is echoed later in *The Rose of England*.

It will now be well to analyse the main sources from which Child drew his collection, distinguishing, as far as possible, between English and Scottish contributions, and to try and disentangle the various types of balladry which are there represented. Most of the few examples attributable to the fifteenth century have already been considered. The only exception is *Riddles Wisely Expounded*, which is found in a manuscript of about 1450, with the heading *Inter diabolus et Virgo*, and the narrator's *incipit*:

Wol ye here a wonder thynge
Betwyxt a mayd and the foule fende?

[1] layne, 'disguise it', 'lie'.

It is in fact little more than a series of riddles, the answering of which confounds the adversary. It had a long life. Versions are found in English broadsides and Scottish tradition, and one was still sung in Wiltshire as late as about 1914. Even from the sixteenth century few texts have come down to us, although entries in the *Stationers' Register* and occasional references or quotations in plays suggest that edacious time has robbed us of others. Half a dozen are found in English manuscripts of this century and three in English prints. Of the latter, *The Fair Flower of Northumberland* and *Flodden Field* come from Thomas Deloney's *History of John Winchcombe*, probably written about 1597, and may, if not wholly of his own composition, have undergone some literary sophistication. There are no 'Child' ballads in the Shirburn collection of broadsides, made about 1600–16, although this contains two late examples of the King and Subject theme. A few may be among the tales and songs attributed to shepherds in the *Complaynt of Scotland* (1549). Captain Cox in 1575 knew *Robin Hood* and *Adam Bell* as 'matters of storie', but the seven examples cited from his 'bunch of ballets and songs' are again not of the 'Child' types. It does not look as if there had as yet been any great development of these types in either country. Child's earliest English ballad from a sixteenth-century manuscript is *Crow and Pie*, a humorous account of a rape, with an *incipit* of the *chanson d'aventure* kind, a touch of the minstrel in the narrative, and a varied semi-refrain, which comes not quite regularly in the last line of the quatrain stanza. Possibly as early may be *The Unquiet Grave*, if that is rightly classed as a ballad at all. The full text only comes to us from late English tradition. But it begins:

> The wind doth blow to-day, my love,
>   And a few small drops of rain;
> I never had but one true-love,
>   In cold grave she was lain.

And this looks very much like the following fragment from a song-book of the first decade of the sixteenth century:

> Western wind, when wilt thou blow,
>   The small rain down can rain?
> Christ, if my love were in my arms,
>   And I in my bed again!

More important is *Adam Bell, Clim of the Clough, and William of Cloudesly*. Of this there is a fragmentary print from the press of John Byddell in 1536, and a full one from that of William Copland in 1548–68. It is a long poem of 170 stanzas, on a greenwood theme like that of the early Robin Hood ballads, to which its debt is obvious. And it is again yeoman minstrelsy. 'Mery it was in grene forest', says the *incipit*, where men walk with bows and arrows,

> To ryse the dere out of theyr denne;
> Suche sightes as hath ofte bene sene,
> As by thre yemen of the north countrey,
> By them it is as I meane.

They are outlaws. Then comes the minstrel's characteristic appeal to his audience,

> Now lith and lysten, gentylmen.

William of Cloudesly goes to Carlisle to see his family. An old wife betrays his presence to the justice and sheriff, and is rewarded with a gown,

> Of scarlat it was, as I heard sayne.

William is beset. An arrow from him breaks on the justice's coat. His house is fired. His wife and children escape by a back window. He takes a sword and buckler and runs among the press, but is captured and threatened with hanging. A little boy warns Adam and Clim, who go to the rescue.

> Her is a fyt[1] of Cloudesli,
> And another is for to saye.

Adam and Clim beguile the porter with a pretended message under the King's seal, wring his neck, and take his keys. They shoot the justice and sheriff beneath the gallows, cut William's bonds, and flee with him to Inglyswode.

> Here is a fytte of these wyght[2] yongemen,
> And another I shall you tell.

William's wife makes her way to the forest and joins them.

> And whan they had souped well,
> Certayne withouten leace,[3]
> Clowdysle sayde, We wyll to oure kynge,
> To get vs a chartre of peace.

---

[1] yt, 'division of a poem'.   [2] wyght, 'strong'.   [3] leace, 'falsehood'.

They go to London, enter the king's hall, and tell the usher, with a curious repetition of the minstrel's own phrasing,

> Syr, we be outlawes of the forest,
>   Certayne withouten leace,
> And hyther we be come to our kynge,
>   To get vs a charter of peace.

The king threatens to hang them, but the queen persuades him to leniency, and he bids them go to meat.

> They had not setten but a whyle,
>   Certayne without lesynge,[1]
> There came messengers out of the north,
>   With letters to our kyng.

In these he is told that they had slaughtered the justice, sheriff, and mayor of Carlisle, together with all the constables, catch-polls, bailiffs, beadles, serjeants of law, and forty foresters of the fee.

> So perelous outlawes as they were
>   Walked not by easte nor west.

He sighed and could eat no more, but said he would see them shoot at the butts against his own archers. They are victorious, and William even ties his son to a stake, and shoots an apple from his head, under peril of hanging, if he fails, at six score paces. The whole family are rewarded with posts at court. The yeomen go to Rome to be assoiled of their sins, and die good men all three.

> Thus endeth the lyues of these good yemen,
>   God sende them eternal blysse,
> And all that with hande-bow shoteth,
>   That of heuen they may neuer mysse!

It has become almost a burlesque of Robin Hood. But here is the full quality of the minstrel, with his seasonal *incipit* and request for a hearing, his asseverations of knowledge and good faith, and his transitions from episode to episode of his tale.

Two ballads from English manuscripts of the middle of the sixteenth century deal with an historical theme, the battle of Otterburn, fought in 1388 between the forces of Richard II and Robert II of Scotland. Of this we have accounts in the chronicles of Froissart and Thomas Walsingham. The Scots gathered at Aberdeen, crossed the Cheviots, and raided the

---

[1] lesynge, 'falsehood'.

Border. Most of them threatened Carlisle, but a detachment under James, Earl of Douglas, and others went towards New-castle. Here, in a skirmish, Douglas took the pennon of Sir Henry Percy, known as Hotspur. On his way home he stopped at Otterburn, to give Hotspur a chivalrous chance of recovering it. A battle followed, in which Douglas was slain by three spearmen and Hotspur was taken prisoner, apparently by Sir John Montgomerie, whose son Hugh had fallen. In the first of our ballads, which has an English tone, history is a little slurred over, apparently in the interest of the Percies. Douglas falls to Hotspur's spear, and although the capture of Hotspur is admitted, it is with reluctance:

> Ther the Dowglas lost hys lyffe,
> And the Perssy was lede away.

> Then was ther a Scottysh prisoner tayne,
> Syr Hewe Mongomery was hys name;
> For soth as I yow saye,
> He borowed[1] the Perssy home agayne.

In this ballad we have still sheer minstrelsy. There is, indeed, no initial appeal for an audience, but only a seasonal *incipit*:

> Yt fell abowght the Lamasse tyde,
> Whan husbondes wynnes ther haye,
> The dowghtye Dowglasse bowynd hym to ryde,
> In Ynglond to take a praye.

But the minstrel's asseverations come in seventeen of the seventy stanzas, generally in the form,

> For soth as I yow saye.

But there are variations:

> For soth and sertenlye.

> For soth withowghten naye.

> I tell yow in sertayne.

> I tell you wythowtten drede.

> I tell you in this stounde.[2]

> As I haue tolde yow ryght.

And most revealing of all,

> The cronykle wyll not layne.[3]

---

¹ borowed, 'ransomed'.   ² stounde, 'moment'.   ³ layne, 'lie'.

Characteristic of the minstrel, again, is the pious *explicit*:

> Now let vs all for the Perssy praye
> To Jhesu most of myght,
> To bryng hys sowlle to the blysse of heven,
> For he was a gentyll knyght.

Two late versions of this ballad have come down to us from Scottish tradition. They must be in some way derivative, since they preserve the Lammas *incipit* of the English text. But here the emphasis is laid on the Scottish victory. In one Douglas is slain by a boy of his own company with a little penknife, and Percy is taken by Sir Hugh Montgomery. Nothing is said about borrowing him home again. Here comes the line:

> Then Percy and Montgomery met,

which is adapted as the name of a song in *The Complaynt of Scotland* (1549). Possibly, therefore, there were divergent English and Scottish versions from the beginning. The second traditional Scottish one makes Douglas fall to Percy's sword and receive burial by Montgomery. Here come the beautiful lines, a variant of which so moved Sir Walter Scott:

> I dreamd I saw a battle fought
> Beyond the isle o Sky,
> When lo, a dead man wan the field,
> And I thought that man was I.
>
> My wound is deep, I fain wad sleep,
> Nae mair I'll fighting see;
> Gae lay me in the breaken[1] bush
> That grows on yonder lee.
>
> But tell na ane of my brave men
> That I lye bleeding wan,
> But let the name of Douglas still
> Be shouted in the van.
>
> And bury me here on this lee,
> Beneath the blooming brier,
> And never let a mortal ken
> A kindly Scot lyes here.

A distinct ballad on the same battle of Otterburn is found in a manuscript which belonged to Richard Sheale, the minstrel of Tamworth, and may have been copied by him from a broadside. Child gives it the title *The Hunting of the Cheviot*. In a version

---

[1] breaken (bracken).

found in the Percy MS. and several English and Scottish broadsides, from the seventeenth century onwards, it is *Chevy Chase*. *The Hunting* is even more remote from historical verity than *Otterburn*. The scene is laid in the Cheviot hills, where not Hotspur but Earl Percy goes to hunt, in defiance of Douglas, and the event is put in the reign of Henry IV instead of that of Richard II. Douglas is killed by an arrow, Percy by Sir Hugh Montgomery, Montgomery himself by another arrow. But the battle is called Otterburn. King Henry avenges it in that of Homildon Hill (1402). Here minstrelsy is again apparent in an *explicit* of prayer, and in one characteristic transition:

> That day, that day, that dredfull day!
> The first fit[1] here I fynde;
> And youe wyll here any mor a the hountynge a the Chyviat,
> Yet ys ther mor behynde.

A later version, called *Chevy Chase*, is found in the Percy MS., in several English broadsides of the seventeenth century, and in one Scottish one of the eighteenth. Prayers in the *incipit* and *explicit* alone suggest minstrelsy. This version also may have been known to the author of *The Complaynt of Scotland*, who records a song of *The huntis of cheuet*. One would be glad to know which ballad it was that stirred the sensitive soul of Sir Philip Sidney. The list of songs in the *Complaynt* also includes one on *The battel of the Hayrlau*, but of this we have no contemporary version. One from nineteenth-century Scottish tradition does not look much like minstrelsy. The 'I' of the *incipit* also takes part in the action of the narrative.

The most important document for seventeenth-century balladry is the Percy MS. Of this, and of the editing which its contents appear to have undergone, something has already been said. It contains no less than forty-six ballads, often unfortunately left in fragments by Humphrey Pitt's housemaids, and of these no less than nineteen are not found elsewhere. Several types are represented. There is a second text of *Adam Bell*. There are eight ballads of Robin Hood, but of these only one is unique, the tale of *Guy of Gisborne*, which has already been traced as existing in some form as far back as 1475. That of *Friar Tuck* may also be of early origin. There are six ballads, including four unique ones, the themes of which are taken from medieval romance. There are fourteen, five of them unique,

---

[1] fit, 'division of a poem'.

which can only be described as imaginative. But a main interest of the collector appears to have been in historical ballads, of which there are no less than seventeen, eight unique ones and nine others. Perhaps some of these would be better described as pseudo-historical, or at the most quasihistorical. Of *Sir Aldingar* enough has perhaps been said. The personages of *Hugh Spencer* and *Sir John Butler* existed, but the incidents described in the ballads lack verification. Of the strictly historical ballads *Durham Field*, *Chevy Chase*, *Musselburgh Field*, and *Sir Andrew Barton* describe battles on land and sea between English and Scottish, and are written from an English standpoint. *The Rose of England*, which celebrates the coming of Henry VII and the battle of Bosworth, is on a purely English theme. So is *Thomas Cromwell*. And although the themes of *Earl Bothwell* and *King James and Brown* are Scottish, the tone is still English. The Scots are 'false' and 'cruel', and 'false Scotland' is contrasted with 'merry England'. The subject of *Captain Car*, on the other hand, is an internal Aberdeenshire feud, with which England was not concerned. Three other historical ballads, *The Rising in the North*, *Northumberland Betrayed by Douglas*, and *The Earl of Westmoreland*, are of special interest. They deal with the fortunes of Thomas Percy, Earl of Northumberland, and Charles Neville, Earl of Westmorland, who led a Catholic rebellion against Queen Elizabeth in 1569 and on its failure took refuge in Scotland. Here the sympathies of the balladwriter are wholly with the rebels. The hero of a fourth ballad, *Jock o' the Side*, took part in the adventures of the fugitive earls, but the ballad itself only deals with a Border raid. The tune of it was known in England as early as 1592. I think the inference must be that the historical interest of the compiler of the Percy MS. lay in England rather than in Scotland, but in England of the northern Border, rather than in southern England. And in so far as earlier versions of his ballads can be traced, these come from English sources. Besides *Robin Hood and Guy of Gisborne*, *Robin Hood and Friar Tuck*, *Adam Bell*, and *Chevy Chase*, we have *Sir Andrew Barton*, *Captain Car*, and *King John and the Bishop*, all of which are in English manuscripts of the sixteenth century. The *Stationers' Register* records prints of *Robin Hood and the Pinder of Wakefield* in 1557–8, *The Lord of Lorne* in 1580, and *Little Musgrave and Lady Barnard* in 1630. This last seems to be quoted in Beaumont's *Knight of the Burning Pestle* of about 1607, and *The Rose of England*, at least by title, in Fletcher's *Monsieur Thomas* of about 1616. Another point that may be noted in the Percy ballads is that they tend to be of considerable length. Many of them are of course fragments, but of those which are complete or practically so, over thirty run to more than thirty stanzas, fourteen to over fifty, and two to over a hundred. This is of course unusual in later ballads, and of itself suggests an element of minstrelsy. And of minstrelsy there is some other evidence, although it seems to be rather a survival. A typical example is *Durham Field*, which has the minstrel's characteristic appeal to his audience as *incipit*, his prayer as *explicit*, his asseveration of evidence in 'as I heard say', and much emphasis on yeomanry. *Will Stewart and John* begins with a curious stanza, which has nothing to do with the narrative and looks as if it came from a lyrical poem:

> Adlatt's parke is wyde and broad,
> And grasse growes greene in our countrye;
> Eche man can gett the loue of his ladye,
> But alas, I can gett none of mine!

This is, however, followed by 'I sing my song', and later come four asseverations of veracity in such terms as 'All this is true that I do say'. *The Rose of England* begins with a curious allegory in which a boar roots up the garden and an eagle saves a branch of the rose. Then follows the minstrel's appeal:

> But now is this rose out of England exiled,
> This certaine truth I will not laine;[1]
> But if itt please you to sitt a while,
> I'le tell you how the rose came in againe.

Later we get, twice over, the revealing phrase:

> The chronicles of this will not lye.

The appeal to the audience is also in *The Rising in the North*, *Northumberland Betrayed*, *King Estmere*, and *Robin Hood and Queen Katherine*; the minstrel's transitions in *Young Andrew*, *Tom Potts*, and *The Lord of Lorne*, always in the form 'Let us leave talking', and his asseverations of good faith, again in *The Lord of Lorne*, 'I tell you all in veretie', although elsewhere this is often reduced to a simple 'I-wis' or 'I wott' or 'trulie'. Sometimes, however, the *incipit* is merely seasonal, or a dream, or a *chanson d'aventure*, or the minstrel's characteristic *explicit* of prayer becomes just

[1] laine, 'conceal'.

a moral or patriotic sentiment. Rather exceptional is *Earl Bothwell*. Here the narrator begins with an imprecation of woe on 'false Scotland'. Then he wins the sympathy of his audience with:

> But you haue heard, and so haue I too,
> A man may well by gold to deere,

follows on with 'I shall you tell how itt befell', and winds up on Queen Mary with:

> But shee is ffled into merry England,
> And Scottland to a side hath laine,
> And through the Queene of Englands good grace
> Now in England shee doth remaine.

It is rarely that we find a ballad which gives the date of its own composition so clearly. *Musselburgh Field* dates precisely the event narrated, but with the minstrel's addition 'as I remember'.

Besides the three Robin Hood ballads already referred to, the Percy MS. has five others. Of these other versions exist in broadside form. The outlawed yeoman remained a popular hero in England up to the nineteenth century. A prose life of him, written in the late sixteenth or early seventeenth, is in a Sloane MS. It was followed in 1632 by Martin Parker's *True Tale of Robin Hood*, which may perhaps claim to be something more than a ballad. But of ballads there are about thirty in broadsides and garlands, some earlier than the Percy MS., but most later. One group of eight, in an unusual five-line stanza, is attributed to a T. R., identified by Dr. Rollins as Thomas Robins. Another has the initials of Laurence Price (*c.* 1628–80). With very few exceptions the Robin Hood ballads have, as already noted, a nonsense refrain, in the form of 'With a hey down down and a down', or a slight variant. There is generally a minstrel's *incipit* and often an *explicit*. One ballad, on *Robin Hood and the Beggar*, comes from a Scottish print of the nineteenth century. Robin occurs also, perhaps by confusion, in *Willie and Earl Richard's Daughter*, and in one version of *Rose the Red and White Lily*. It is curious that Scotland does not yield more, in view of the early popularity of Robin as a leader in its folk-revels, and of the inclusion of 'Robene Hude and litil ihone' among the shepherds' tales recorded in *The Complaynt of Scotland* (1549). It is curious that, while only four of the Percy ballads, other than those on Robin Hood, have come down to us in later versions from English sources, no less than eleven are traceable

in Scottish tradition. This may perhaps be due to the copying
of English broadsides, now lost, by Scottish printers. But in
fact the total English contribution, later than the date of the
Percy MS., to Child's collection is a comparatively small one,
amounting only to thirty-five ballads, of which twenty-seven
come from broadsides and other prints, and eight from oral
tradition. No doubt this number could now be increased
through recent findings among folk-singers. Mr. Williams has
in fact already unearthed a version of the *Sir Lionel* of the
Percy MS., as *Bold Sir Rylas*, in Wiltshire. Of Child's thirty-
five twenty-two must be classed as imaginative and thirteen
as historical or at least quasihistorical. Three of the thirteen,
Rookhope Ryde, *Hughie Grahame*, and the admirable *Death of Parcy
Reed*, describe Border forays. Scotland was, of course, at least
as interested as England in Border happenings. The Crosiers
who slew Parcy came in fact from Liddesdale. But it is again
noteworthy that nineteen of the thirty-five English ballads also
survive in Scottish tradition. In English ballads, minstrelsy is
generally recessive, amounting to little more than an occasional
incipit or *explicit*, or both of these. But sometimes the *incipit* is
merely seasonal, or records an incident, seen 'as I lay musing'.
Bewick and Graham has, however, in two places, the familiar
'Let us leave talking', and *Rookhope Ryde* is full of the minstrel's
'I wat' and 'I trow'.

It is now time to turn to the large group of ballads which
may reasonably be regarded as belonging to Scottish tradition,
even though some of them are also found in English versions.
They are about 180 in number, thus making up about two-
thirds of the total represented in Child's volumes. And they
are of the first importance, not merely for their bulk, but for
their literary quality, since they include most of those which,
if the anthologists may be trusted, must be regarded as the
best ballads. Here are, for example, such admirable things as
Sir Patrick Spens, *Sir Hugh*, *Johnie Cock*, *Marie Hamilton*, *Clerk
Saunders*, *Edward*, *Lord Randal*, *Thomas Rymer*, and *Tam Lin*.
Many others might be added. The problem of the origin of
Scottish balladry must be deferred for the present. Unfortunately we are dependent for the traditional material upon very
late sources. Collecting in Scotland may be said to have begun
in the first half of the eighteenth century with the *Tea Table
Miscellany* of Allan Ramsay already recorded. Some early

additions were made by other Scottish booksellers and printers. Bishop Percy obtained others from various correspondents, which he utilized for his *Reliques* (1765). A vigorous period of research followed, in which a large number of ballads were recovered from singing or recitation by David Herd (1769, 1776), John Pinkerton (1783, 1786), Joseph Ritson (1785–1821, 1794), Sir Walter Scott (1802–3), Robert Jamieson (1806), John Finlay (1808), Robert Hartley Cromek (1810), Alexander Laing (1822, 1823), Charles Kirkpatrick Sharpe (1823), James Maidment (1824, 1828), Peter Buchan (1825, 1828), Allan Cunningham (1825), Robert Chambers (1826, 1844), William Motherwell (1827), George Ritchie Kinloch (1827). Many additions have, of course, been made later, notably since Child's day by Mr. Gavin Greig, who found a number of new ballad-versions, although apparently no completely new ballads, in Aberdeenshire.

Of the 180 ballads, by far the greater number must be classed as purely imaginative. About forty have claims to be considered as historical. In several cases, however, quasihistorical would be the safer term. This is certainly so with two, the incidents of which have been ascribed to the thirteenth century. *Hugh of Lincoln* rests upon a legend which is in fact English and not Scottish. In some versions of the ballad Lincoln has become a mythical Mirryland. The shipwreck in *Sir Patrick Spens* has been conjecturally related to disastrous incidents in voyages from Norway to Scotland. The records of these yield no such name as Spens. But the ballad-writer had a fine touch on wonder and pathos:

> Late late yestreen I saw the new moone,
>   Wi the auld moone in hir arme,
> And I feir, I feir, my deir master,
>   That we will cum to harme.

And again:

> O lang, lang may their ladies sit,
>   Wi thair fans into their hand,
> Or eir they se Sir Patrick Spence
>   Cum sailing to the land.

> O lang, lang may the ladies stand,
>   Wi thair gold kems in their hair,
> Waiting for thair ain deir lords,
>   For they'll se thame na mair.

Half a dozen ballads are on themes of outlawry and Border raids and forays, much like those of English origin. The most interesting is *Johnie Cock*, which has some curious primitive touches. There are wolves about, and men drink the blood of the slain deer. There are ballads which celebrate battles. Of that on the early-fifteenth-century *Battle of Harlaw* enough has already been said. Four others describe conflicts between royalists and covenanters in the seventeenth century. There is little treatment of outstanding political events. The intrigue of the Earl of Bothwell and his adherents against James VI in 1592, and the execution of the Earl of Derwentwater, who rose for the Pretender in 1715, are exceptions. Four ballads record murders. The most interesting is *Marie Hamilton*. A rather eccentric attempt has been made to find the origin of this in Russian instead of Scottish history. No such Mary is in fact recorded among the four who served the Queen of Scots in 1563, but the ballad makes her one of them, executed for drowning a child she had borne to 'the hichest Stewart of a' '.

> Last nicht there was four Maries,
> The nicht there'l be but three;
> There was Marie Seton, and Mary Beton,
> And Marie Carmichael, and me.

Several ballads deal with feuds between Scottish families, and many with domestic events, murders again, seductions and elopements, marital complications, a death through plague. The issues are generally, although not always, tragic. These domestic ballads cannot be sharply differentiated from those of the imaginative group, which are also apt to use the names of known families. And again some of them can only be regarded as quasihistorical, since one may suspect that they are themselves often the sources of the corroborations quoted in their support. Chronologically the feuds and domesticities concerned belong, when they can be dated, to the sixteenth and still more the seventeenth century. Two are of the eighteenth. One is ascribed in an early family chronicle to the fourteenth. Some of the historical ballads still contain touches of minstrelsy, in *incipit* or *explicit*, or in interspersed phrases such as 'I wat' or 'quo he'. They are commonest in the Border and battle groups. A seasonal *incipit* is occasionally substituted.

The Scottish imaginative ballads mostly come, not from the Border, but from Aberdeenshire in the north. This was a centre of maritime commerce, and there is often an element of sea-faring in the stories. A few ballads take their themes from earlier romance. A few others are merely humorous, with riddles and the like. One only is religious, on a miracle of the Virgin. But the predominant interest, to a somewhat surprising extent, is sexual. Here the Scottish imagination seems to be in revolt against the austerity of Scottish religion. It is the woman, rather than the man, who is in the forefront of the picture. In a very large number of cases the action begins with a seduc-tion, in the lady's bower, or, it may be, in the greenwood, where she has gone to pull a nut. An elopement, with a view to marriage, may, or may not, follow. Occasionally the seduc-tion is complicated by incest, conscious or unconscious. But if the starting-point is a commonplace, there is room for much variety in the final outcome. Examples only can be given here. It is often tragic. The incestuous girl must of course die, by her own hand or her brother's. He may commit suicide, or go off in a bottomless boat. In other cases the seducer is killed, or brought to hanging, by the girl's relatives. She dies, or goes mad, or at least mourns for seven years. He may prove faith-less, and she kills him or is killed by him. He was after her money and she leaves him. He commits a murder; she betrays him and regrets it. She is forced to marry another, and he kills himself. We do not know what lies behind the story of Lord Randal, whose lady poisoned him with a broth of what he thought were eels, but must in fact have been snakes. But true love, as well as seduction, may end in tragedy. The ill will or pride of relatives leads to murder before a wedding day. A girl is forced by her parents to marry the wrong man. She rejects a lover in a fit of pique, and two hearts break. A bride-groom is smitten with death as he rides away with his bride. When true lovers die together it is usual to plant a rose brier on either grave, and as they grow the briers twine together and become one. This pretty fancy is found in English ballads also. Marriage itself is of course no safeguard against tragedy. A husband is unfaithful, or there is a marital quarrel and the wife dies. He kills his wife's brother; she saves him from pursuit, but he kills her, too. A wife poisons a former lover, who rejected her, and has now become the lover of another woman. Only a very few tragic ballads lack a sexual element. A man murders his brother at his mother's instigation. A rebel kills a king, but

the king's son avenges him. An unpaid mason kills the wife and
son of his employer and is hanged. The story is said to have
frightened northern nurseries for generations. But the Scottish
mind is attuned to sentiment as well as to tragedy. About half
the ballads, whether they start with a seduction or not, are
romantic, with a happy ending. Lovers, especially women,
prove faithful, and the reward is often marriage with a husband,
who turns out unexpectedly to be a nobleman. No doubt there
are obstacles to be overcome first. There has sometimes been
a temporary alienation, or a hazardous voyage at sea. A wicked
step-mother causes trouble. Other hostile relatives or rivals
have to be defeated, and perhaps slain. The girl must be
rescued from abduction by another or taken from a nunnery. A
false accusation has to be repelled or a crime condoned. A
princess cannot wed a subject without the hardly won consent
of her father. But in the end the marriage bells ring, and all is
well. If, on the other hand, marriage is itself the beginning of
the story, there are conjugal quarrels to be composed, or conjugal infidelity to be forgiven. Romantic, as well as tragic,
ballads, are rarely without a sexual element. But there are two
which turn upon a conflict between a farmer or a merchant
and a highwayman, and in both the honest traveller gets the
better of the thief. In a third, a seaman is alone saved from the
shipwreck of a fleet, by the help of Bonny Boy.

There is often a strong element of the supernatural. Similar
themes can be traced elsewhere in European literature and
folk-lore. Ghosts are abroad. A 'griesly' one makes a man
seduce her, and in the morning she is a fair lady. A girl goes
with another, but it will be to clay. A third is that of a dead
husband, who promises to be his wife's porter at heaven's gate.
In *The Wife of Usher's Well*,

> It fell about the Martinmass,
>   When nights are lang and mirk,
> The carlin wife's three sons came hame,
>   And their hats were o the birk.[1]
>
> It neither grew in syke[2] nor ditch,
>   Nor yet in ony sheugh;[3]
> But at the gates o Paradise,
>   That birk grew fair eneugh.

---

[1] birk (birch).   [2] syke, 'stream'.   [3] sheugh, 'trench'.

They must be gone at daybreak.

> Fare ye weel, my mother dear!
> Fareweel to barn and byre!
> And fare ye weel, the bonny lass
> That kindles my mother's fire!

A helpful spirit is known as Belly Blind. In one ballad he unites parted lovers, and in another counteracts the charm wrought by a hostile mother-in-law to prevent childbirth. Drops of St. Paul's blood act as a charm to revive the dead. There are transformations. A girl thrown into the sea by her step-mother becomes a savage beast, but resumes her natural shape under the kisses of her lover. A witch turns a man who rejected her into a worm, but the fairy queen restores him on Hallow E'en. A supernatural lover goes to church and gets Christendom. A ship is prevented from travelling by fey[1] folk, until a seduced girl is thrown overboard. Talking birds bring messages or warnings to lovers. The stone of a magic ring fades on a death. Another ring enables a man to kill a three-headed giant. Of particular interest are ballads which recount visits to fairyland. The chief of these is *Thomas Rymer*. It is based on a romance, probably written towards the end of the fourteenth century, which incorporates prophecies, traditionally ascribed to one Thomas of Ersseldoune. In the ballad the Queen of Elfland takes him on her milk-white steed.

> For forty days and forty nights
> He wade thro red blude to the knee,
> And he saw neither sun nor moon,
> But heard the roaring of the sea.

The story takes at one point a religious turn, unusual in ballads. Thomas is shown the narrow road, beset with thorns and briers, which is the path of righteousness, and the broad road over a lilied field, which is the path of wickedness. But neither of these is for him. He is bound elsewhere.

> But Thomas, ye maun hold your tongue,
> Whatever you may hear or see,
> For gin ae word you should chance to speak,
> You will neer get back to your ain countrie.

---

[1] fey, 'fatal', 'supernatural'.

He has gotten a coat of the even[1] cloth,
And a pair of shoes of velvet green,
And till seven years were past and gone
True Thomas on earth was never seen.

A variant story is that of *Tam Lin*, who is a danger to maidens haunting Carterhaugh with gold on their hair. It may be that known to the author of the *Complaynt of Scotland* as early as 1549. He had been taken by the Queen o' Fairies, and became an 'elfin grey'.

And pleasant is the fairy land,
But, an eerie tale to tell,
Ay at the end of seven years
We pay a tiend[2] to hell;
I am sae fair and fu o flesh,
I'm feard it be mysel.

Fair Janet admires him, and on Hallow E'en holds him in her arms while he is turned successively into an esk,[3] an adder, a bear, a lion, and 'a red hot gaud[4] of iron'. The spell is broken. One unusual ballad comes, not from Scotland proper, but from Celtic Shetland. A woman has a child by a great silkie.[5] He takes it and will teach it to swim, but prophesies that some day the woman will marry a gunner, who will shoot both father and son.

Minstrelsy, in the imaginative ballads, may be said to have become merely vestigial, little more than the recognition of a tradition from the past, which has ceased to be effective. The singer no longer makes his direct appeal for an audience. At the most he may refer to his coming story as something he has seen or heard, or give advice to keep away from the dangerous greenwood. If there is a formal *incipit*, it is seasonal or adventurous. An *explicit* is rare. Occasionally one offers a comment:

But it would have made your heart right sair
To see the bridegroom rive his hair.

Or:

An I hope ilk ane sal sae be served,
That treats an honest man sae.

But as a rule, the narrative plunges at once *in medias res*, and ends as abruptly. '*Ballad* is an Idea, a poetical Form', says Professor Ker, 'which can take up any matter, and does not

---

[1] even, 'smooth'. [2] tiend, 'tithe'. [3] esk, 'newt'.
[4] gaud, 'bar'. [5] silkie, 'seal'.

leave the matter as it was before', and Professor Entwistle echoes him. Perhaps that does not carry us very far, since it is true of other literary types, as well as the ballad. Professor Gerould has made a valuable study of ballad form, although it applies more directly to the Scottish examples than to the earlier ones of English origin. He notes three 'constants'. One cannot do better than quote him:

One of these constants is stress on situation, rather than on continuity of narrative or on character as character is presented in heroic poems or prose sagas. Not only are all ballads stories of action, but they are stories in which the action is focussed on a single episode. Sometimes, to be sure, a whole series of events in the past is revealed by the incident which is the subject of the narrative, but only by reference. . . . More commonly the past is ignored altogether, or is implied rather darkly, and the situation is presented for itself alone.

This feature Professor Gerould finds characteristic of continental as well as English and Scottish ballads. And so, too, with his second constant, which is closely related to the first:

Whatever the matter of a ballad may be, and whatever the manner of presentation in other respects, there is always a marked tendency to tell the story dramatically. The brevity of the narrative has much to do with this, no doubt, but it does not explain everything, since a short poem may well be entirely without dramatic quality. Ballads, however, are not merely short: they are compressed. The series of events is seized at its culminating point and is envisaged in terms of the action which then takes place; nothing matters except the action; the characters speak because they have thoughts to express about what is taking place. Little as the dialogue of ballads has to do with the talk of actual life, it has at its best a trenchant pertinency not to be matched except in highly developed drama.

Professor Gerould finds his third constant perhaps less completely typical of balladry outside these islands:

This is the impersonal attitude to the events of the story that is at least the rule among ballad-makers. The story is told for the story's own sake, while the prepossessions and judgements of the author or authors are kept for the most part in the background.

It is a helpful analysis. Two other stylistic points may be noted, which Professor Gerould does not overlook. One is the frequent use of incremental repetition, a feature which is also often found

in carols and other forms of medieval poetry related to carols.
The other, closely analogous to the first, is the reliance upon
commonplace stanzas, which recur, with little variation, in
ballad after ballad. Must a lady ride,

> 'Gar sadle me the black', she sayes,
>   Gar sadle me the broun;
> Gar sadle me the swiftest steed
>   That ever rode the toun.

Or, when a little foot-page is sent to take a love-letter or give
a warning,

> O whan he came to broken briggs,
>   He bent his bow and swam,
> An whan he came to the green grass growin,
>   He slack'd his shoone and ran.

There is a stern economy here in the reduction of the unessential
to a formula.

We approach the question of the origin of balladry, and in
particular of Scottish imaginative balladry, upon which the
attention of scholars has perhaps been rather unduly concen-
trated. The earlier northern collectors, such as Sir Walter Scott
and William Motherwell, were content to regard it as a final
outgrowth of minstrelsy. But the growing interest, during the
last half of the nineteenth century, in the 'folk' and its ways of
life and thought led to a revival of the old German theory of
Das Volk dichtet. A protagonist was Andrew Lang, himself a
learned folk-lorist, who set out his views, with much poetic
enthusiasm, in the *Encyclopaedia Britannica* (1875):

Ballads sprang from the very heart of the people, and flit from
age to age, from life to life, of shepherds, peasants, nurses, of all that
continues nearest to the natural state of man. They make music
with the flash of the fisherman's oar, with the hum of the spinning-
wheel, and keep time with the step of the ploughman as he drives
his team. The whole soul of the peasant class breathes in their
burdens, as the great sea resounds in the shells cast up from its
shores. Ballads are a voice from secret places, from silent places,
and old times long dead. It is natural to conclude that our ballads
too were first improvised and circulated in rustic dances.

This is perhaps sufficiently answered by the quiet irony of
George Meredith in *The Amazing Marriage* (1895), where Dame
Gossip's 'notion of a ballad is, that it grows like mushrooms
from a scuffle of feet on grass overnight'. But in the meantime

Lang's enthusiasm had become rather subdued. In his preface to a selection of ballads for Ward's *The English Poets* (1887) he says no more than:

About the authors of the ballads, and their historical date, we know nothing. Like the *Volkslieder* of other European countries, the popular poems of England were composed by the people for the people.

And in an article of 1904 he seems to have abandoned his theory of communal origin altogether. But the doctrine which he expounded with such fervour in 1875 and so discreetly let drop, found a congenial home later in democratic America. Child himself, during the progress of his work, seems to have been rather non-committal. In a note at one point in his collection he distinguishes between the minstrel and the popular ballad. And of the latter he had written in his tentative essay of 1893 that it was 'anterior to the poetry of art, to which it has formed a step, and by which it has been regularly displaced, and in some cases all but extinguished'. And he goes on:

The condition of society, in which a truly national or popular poetry appears, explains the character of such poetry. It is a condition in which the people are not divided by political organisation and book culture into markedly distinct classes, in which, consequently, there is such community of ideas and feelings that the whole people form one individual. Such poetry, accordingly, while it is in its essence an expression of our common human nature, and so of universal and indestructible interest, will in each case be differentiated by circumstances and idiosyncrasy. On the other hand, it will always be an expression of the mind and heart of the people as an individual and never of the personality of individual men. The fundamental characteristic of popular ballads is, therefore, the absence of subjectivity and of self-consciousness. Though they do not write themselves, as William Grimm has said, though a man and not a people has composed them, still the author counts for nothing, and it is not by accident, but with the best reasons that they have come down to us anonymous.

There is some inconsistency here. Idiosyncrasy is allowed for at one point, only to be rejected in what follows. And in the same essay we are bidden to remember,

That tales and songs were the chief social amusements of all classes of people in all the nations of Europe during the Middle Ages, and that new stories would be eagerly sought for by those whose business it was to furnish this amusement, and be rapidly spread among the fraternity.

This fraternity, of course, can only be that of the minstrels.

Professor Kittredge does not perhaps go much farther than Child himself. For him, too, ballads, 'the poetry of the folk', are to be distinguished from 'the poetry of art'.

They belonged, in the first instance, to the whole people, at a time when there were no formal divisions of literate and illiterate; when the intellectual interests of all were substantially identical, from the king to the peasant.

I am not quite sure that any such 'homogeneous folk', as Professor Kittredge elsewhere calls it, ever existed in England, from the time when the Saxons, already well differentiated, alike on military and agricultural lines, entered these islands. But Kittredge may have been influenced by an even more ardent believer in *Das Volk dichtet*, Professor Gummere of Haverford, also a pupil of Child's. Professor Gummere has written much upon ballads, over a term of years. As I understand his latest utterance, it amounts to this. Every ballad was of course composed, as any other poem is composed, by the rhythmic and imaginative efforts of a human mind. But ethnological research has shown us that among peoples in a primitive stage of civilization, before the distinction between a leader and his followers had fully developed, singing was often closely related to a choric dance. This might take place after a successful battle or foray or labour in the harvest field, or in festal mirth at a wedding or sorrow at a funeral. And in the course of it first one and then another might slip out of the throng and improvise his song of exultation or lament over the event celebrated. It would be largely lyrical, no doubt, but some element of narrative retrospection might naturally find a place. As each finished his contribution, the rest of the company would echo his words in a sympathetic cry of triumph or regret. And so, in the excitement of saltation, as one of Professor Gummere's critics has put it, the ballad, with its characteristic elements of iteration and refrain, was born. And in this sense its composition may be called communal. Professor Gummere thinks that this practice of combined dance and song continued to prevail in these islands long after the Saxon invasion, but that conditions favourable to its existence died out in the fifteenth century. Ballads continued to be composed in isolated rural communities, but were only 'a heritage of the past'. Literary influences had worked upon them, and in the process of oral transmission they had lost their dramatic or mimetic and choral character and become 'distinctly epic'. They had even forfeited their 'once indispensable' refrain. No ballads of the earliest type are left to us. Even the Robin Hood ballads are 'far gone in the epic process'. Ballads can be written no longer. 'That merry art is dead.' There is of course much conjecture in all this. Of his supposed medieval ballads Professor Gummere can only find what he thinks one clear trace. That is in the famous song attributed to Canute (*ob.* 1035) in the *Historia Eliensis*. It will be as well to give the record in its original form.

Quodam igitur tempore, cum idem rex Canutus ad Ely navigio tenderet, comitante illum regina sua Emma et optimatibus regni, volens illic, juxta morem, purificationem S. Marie solempniter agere, quando abbates Ely, suo ordine incipientes, ministerium in regis curia habere solent; et dum terrae approximarent, rex in medio virorum erigens se nautis innuit ad portum Pusillum ocius tendere, et tardius navem ineundo pertrahere jubet, ipse oculos in altum contra ecclesiam, quae haud prope eminet in ipso rupis vertice sita, vocem undique dulcedinis resonare sensit, et erectis auribus quo magis accedit amplius melodiam haurire coepit; percepit namque hos esse monachos in coenobio psallentes et clare divinas horas modulantes, caeteros qui aderant in navibus per circuitum ad se venire, et secum jubilando canere exhortabatur; ipsemet ore proprio jocunditatem cordis exprimens, cantilenam his verbis Anglice composuit, dicens, cujus exordium sic continetur:

> Merie sungen the muneches binnen Ely,
>   Tha Cnut ching reu ther by;
> Roweth, cnites, noer the land,
>   And here we thes muneches sæng.

Quod Latine sonat: 'Dulce cantaverunt monachi in Ely, dum Canutus rex navigaret prope ibi, nunc milites navigate propius ad terram, et simul audiamus monachorum harmoniam', et caetera quae sequuntur, quae usque hodie in choris publice cantantur et in proverbiis memorantur.

It is an interesting example, if the record, nearly a century after the event, can be trusted, of improvisation by the leader of a singing crowd. But, as with other early *cantilenae* of which we hear, there is not the slightest evidence that the song was in the narrative form of ballad. It may just as well have been a lyric in praise of the singing of the monks themselves, or of the beauty of the building and its site, or of Canute's own generosity in contributing to its endowment.

Professor Gerould is a cautious historian, and his interpretation of *Das Volk dichtet* is somewhat different from that of Professor Gummere. He does not see any advantage in continuing 'the intermittent warfare that has been carried on for more than a century by communalists and individualists'. We have to consider two distinct problems—on the one hand, 'what gave rise to the mould or pattern of ballads', and on the other, 'how and when the individual ballads of our traditional store were made in accordance with that pattern'. From ethnological evidence we learn,

(1) that the power or the habit of verse-making, though not universal, is more widely diffused among folk with a simple culture than among people whom we call civilized; (2) that songs are ordinarily made as the result of some immediate and definite stimulus, which is more often than not concerned with tribal matters and sometimes results in improvisation; and (3) that song is intimately related to the dance.

But we cannot assume the direct continuance of such primitive conditions of composition into the later Middle Ages, earlier than which we cannot trace the ballad. There is no ground for the belief that the people of northern Europe, before the migrations, chanted anything like it. *Beowulf* and *Waldere* represent a very dissimilar type of narrative poetry. The most we can do is to point to a 'remarkable similarity between the habits of verse-making among uncivilised races and among those large majorities of civilised folk who have not fallen until of late under the immediate influence of schools and the traditions of conscious artistry'.

This may at least help to explain some of the qualities of the ballads, as we have them. Professor Gerould finds the earliest European record of a ballad in the combination of dance and song and story described in the eleventh- or twelfth-century legend of the Dancers of Kölbigk. And for him the important thing here is not so much the association of story with dance as of story with dance tune:

It seems to me that the singing of a narrative to a melody is the kernel of the whole matter. Wherever and whenever that adaptation was made, the ballad as we know it came into being.

He does not think that this development can have taken place under the conditions of communal composition. 'Neither a melody nor the outline of an imagined story can well emerge from more than a single mind.' Individuals, therefore, must have 'fashioned the earliest ballads—those that ultimately set the form'. And some of these may well have been minstrels, of a sort. Where then does the folk come in? Not at the original creation of a ballad, but at its recreation through the process of oral transmission. Cecil Sharp pointed out that it is very rarely that two folk-singers will be found to sing the same song in precisely the same form. He believed

> that the most typical qualities of the folk-song have been laboriously acquired during its journey down the ages, in the course of which its individual angles and irregularities have been rubbed and smoothed away, just as the pebble on the seashore has been rounded by the action of the waves; that the suggestions, unconsciously made by individual singers, have at every stage of the evolution of the folk-song been weighed and tested by the community, and accepted or rejected by their verdict; and that the life-history of the folk-song has been one of continuous growth and development, always tending to approximate to a form which should be at once congenial to the taste of the community, and expressive of its feelings, aspirations, and ideals.

It is, no doubt, through dwelling on the process of oral transmission that Professor Gerould brings himself to say, at the beginning of his book, that the popular ballad, or the ballad of tradition as he prefers to call it, 'has no real existence save when held in memory and sung by those who have learned it from the lips of others'. At any rate, we now become able to attach some sort of meaning to the phrase *Das Volk dichtet*.

Professor Entwistle's account of the effect of oral transmission is much the same as Professor Gerould's:

> Composed in common form, the ballad becomes at once common property, like a fairy-tale or legend. The author has no copyright, and the ballad only exists by virtue of each successive performance when it is what the performer makes it. It is not that ballads were, as the Romantics insisted, the product of the community working as a creator. Artistic creation under such conditions would be impossible; each ballad has its author and its moment of birth.

But 'Once launched, the ballad is everybody's possession'. I must confess that I feel a touch of eighteenth-century Teutonic mysticism still hanging about the emphasis laid in these theories upon the importance of oral transmission. It has of course been a *vera causa* in bringing about the condition in which most ballads have reached us, and one need not quarrel with the statement of Menéndez Pidal that at any moment of the process beauty may drop in. It is, however, also true that it has often worked for degeneration. Mr. T. F. Henderson perhaps goes rather far when he says of Child's collection that 'the chaff is out of all proportion to the wheat'. But there are many mean and some barely intelligible elements in the ballads as they have come down to us, both in structure and in phrasing, for which oral transmission must at least bear its share of the discredit. Sometimes a failure to understand the older forms of a changing language has been at the root of the trouble. But ignorance will not explain everything. The minds of the transmitters have not always been on the same level as those of the original authors. Mr. Keith notes 'the vulgar frills, the tinsel imagery, the clownish expressions' with which we are often confronted. I am not sure that, at its best, oral transmission is quite sufficient to account for all the wider differences between the extant 'versions' of a ballad, as distinct from 'variants' in texts of what is recognizably the same ballad. One may suspect that sometimes a more wholesale rehandling may have taken place. And the attitude of Professor Gerould and others seems to be rather hard upon the original makers of the ballads, who after all must have contributed something. Here, unfortunately, we are very much in the dark. Scottish texts come to us much later than the English Robin Hood ones, with their evidence of yeoman minstrelsy. We can hardly rely upon John Barbour's lines, written in 1375 on a Scottish victory earlier in the century.

> I will nocht reherss all the maner;
> For quha sa likis, thai may heir
> Young women quhen thai will play,
> Syng it emang thame ilke day.

This suggests an element of narrative, but we cannot assume that it was in a form anything like that of a ballad. And the song on Bannockburn (1314) in the fifteenth-century *Brut* was clearly, from its 'rombelow' refrain, a boating one, although Robert Fabyan in 1516 says that it was sung 'after many dayes' by maidens and minstrels in dance and carol. A minstrel, one would suppose, was more likely to have written it than a maiden. The first clear record of Scottish ballads is in *The*

Complaynt of Scotland (1549). The dialect is that of the southern part of the country. The author, in spite of various speculations, remains unknown. He describes a walk, in the course of which he comes to a field where shepherds are sitting down to break-fast. After an oration by one of them, they decide to amuse themselves with 'joyous comonyng'. Then comes, on paper different from that of the rest of the manuscript, a list of their tales, songs, and dances, which are declared to be only a selection, 'sa mony as my ingyne[1] can put in memorie'. It would take not a day, but a month, to go through, and one may suspect that it amounts to a fairly complete catalogue of such literature as was known to the writer. Certainly it includes English as well as Scottish material. In fact it begins with 'the taylis of cantirberrye'. English, again, must be, at least by origin, the tale of 'robene hude and litil ihone'. That 'quhou the king of est mure land mareit the kyngis dochtir of vest mure land' may be a Scottish version of the ballad of *King Estmere*, of which an English one comes from the Percy MS. That of 'the yong tamlene, and of the bald braband' suggests the Scottish *Tam Lin*, although there is no 'bald braband' in Child's texts of that. The songs include 'Brume, brume on hil', possibly related to Child's *The Broomfield Hill*, or alternatively to a refrain sung in Wager's play of *The Longer Thou Livest* (*c.* 1568); also 'The frog cam to the myl dur', probably *A Wedding of the Frog and Mouse*, which was entered on the English *Stationers' Register* in 1580, printed in Thomas Ravenscroft's *Melismata* (1611), and sung on the Scottish stage in the eighteenth century. The entertainment was obviously of a very miscellaneous character. But at least the songs of 'the battel of the hayrlau', 'the hunttis of cheuet', and 'the perssee and the mongumrye met' may reasonably be taken to be versions of ballads that survive. The dances are of 'Robene hude', of 'Thom of lyn', which again, if not related to *Tam Lin*, might be the 'Tom a Lin' of Wager's play, and of 'Ihonne ermistrangis dance', the relation of which to Child's *Johnie Armstrong* is not clear. John Armstrong was a Border reiver, executed in 1530. Possibly his 'dance' was his hanging. One may perhaps infer from *The Complaynt* that in Scotland ballads of warfare and Border foray came earlier than those of the imaginative type. Some confirmation of this may be obtained from John Leslie's account of the Border folk in

[1] ingyne, 'intellect'.

his *De Origine, Moribus, et Rebus Gestis Scotorum* (1578), where he tells of 'cantiones, quas de majorum gestis, aut ingeniosis praedandi precandive stratagematis, ipsi confingunt'. But he gives no names, unfortunately. Mr. T. F. Henderson suggests that ballads may have been written by some of the Scottish 'makaris'[1], who became numerous at the turn between the fifteenth and sixteenth centuries. That is a plausible conjecture, but it can hardly be more. Robert Henryson and William Dunbar, although much under the Chaucerian influence, occasionally approach the ballad manner in their simpler poems. But, even here, their metres tend to be rather more elaborate. There were, however, many others, of whom we know little or nothing. Dunbar gives the names of some of them in his *Lament for the Makaris, quhen he wes seik.*

Scotland seems to have made little, if any, contribution to the Percy MS. of about 1650, although some of its numbers recur in later versions. For the imaginative ballads we have to rely almost entirely upon the collectors, who began their work in the eighteenth century. They, too, learnt nothing of the original authorship. There were, by this time, a few broadsides and chapbooks. But the main gathering was from oral tradition. The conditions in the eighteenth century were probably not very different from those which confronted Mr. Gavin Greig in Aberdeenshire, from the last quarter of the nineteenth century onwards. They are fully described by his editor, Mr. Alexander Keith, who is not much impressed by Teutonic speculations of the *Das Volk dichtet* type. 'Water that's drumlie is nae aye deep.' Greig's versions came mostly from 'rural corners tucked away safely from the chief intellectual amenities of civilisation'. In villages the contributors belonged to what may be called the upper stratum of the peasantry. They were largely farmers, but also crofters, ploughmen, shepherds, drovers, harvesters, and the like. In small towns they were tailors, shoemakers, coopers, spinners, or occasionally they were ministers and schoolmasters, in touch with these. There were a few travelling singers and fiddlers, who hawked ballads from door to door. An earlier example of this type had been James Rankin, whose reputation and that of his employer, Peter Buchan, for faking their texts have been cleared to some extent by Mr. Keith. There seems to have been no dancing of ballads in Greig's day,

---

[1] makaris (makers), 'poets'.

or indeed in that of the older collectors, but girls of an earlier
generation than Greig's would occasionally gather with their
knitting for the singing of one in chorus with a refrain. Some
of the ballads were of poor quality. Mr. Keith gives one, which
'never came from the sources out of which the great ballads
arose', but was none the less 'full of traditional characteristics'.
Others, again, must have been 'the work of poets either of local
or of national reputation or deserving of such fame'. Greig's
chief contributor was Miss Bell Robertson, who was born on
a croft in 1841, received her education from an old woman who
lived by singing, and became a housekeeper. Her ballads came
through her mother, from a grandmother who was a folk-singer.
Mrs. Brown of Falkland, on the other hand, whose ballads
Scott used, was an educated woman, the daughter of an Aberdeen professor and the wife of a clergyman. She was born in
1747, and acquired her repertory by her twelfth year from her
mother and aunt and her mother's maid. How often, throughout the ages, has not a domestic servant been an intermediary
between the 'folk' and the gentry?

One other aspect of the imaginative ballads requires consideration. They have come down to us through the peasantry,
but the personages of their stories are not as a rule peasants.
On the contrary, they are predominantly high-born. Here are
kings' daughters and earls, even if in disguise. Knights are
as plentiful as blackberries. The action is often in a castle. At
the worst, the daughter of the house has her bower to be seduced
in. Servants are to hand, including the invaluable little foot-
page. There is much gold and silver about. The horses are
shod with it. Somewhere, therefore, in the ancestry of these
particular ballads there is a courtly strain, which the fancy of
a lower class has appreciated and elaborated. It may be related
to the fact that the themes of the imaginative ballads often have
close parallels in Scandinavian poetry and legend. Professor
Entwistle, in his learned survey of European balladries, regards
those of our islands, with those of Scandinavia and Germany,
as forming a distinct Nordic group. German ballads were a
comparatively late development. Those of Scandinavia began
in Denmark and spread to Norway, Sweden, Iceland, and the
Faroese islands. Through sea-traffic they made their way to
England and Scotland. He does not think it necessary to ascribe
this process to the days of Canute's empire and the Danelaw.

And indeed we need not even ascribe it to the Middle Ages, since commerce over the North Sea was still active in the fifteenth and sixteenth centuries. Danish balladry itself, he thinks, may have owed something to French influences. This point is dealt with more fully by Professor W. P. Ker, who touched no topic of literature without illuminating it. He believes that in the twelfth century the French dancing song, known as the *carole*, made its way into Denmark and was there fitted to stories taken from earlier Scandinavian sources. To the *carole* we must add, I think, the more definitely narrative *chanson de toile*, which M. Jeanroy believes to have been danced, as well as sung at needlework. Certainly the lines attributed to the dancers of Kölbigk look more like those of a *chanson de toile* than of a *carole* proper. In Denmark, says Professor Ker, the ballads were originally the entertainment, not of the people as a whole, but of the gentry—'a gentry not absolutely cut off nor far removed from the simpler yeomen', and they remained in favour among the ladies of the country down to at least the seventeenth century. If then the stories of ballads came across the sea from Scandinavia, the aristocratic tinge of the Scottish versions becomes intelligible. And there is at least one feature of them which may even point to an ultimate French origin. 'Bele Aelis' is the heroine of a number of *caroles*. The *chansons de toile* generally begin with a 'Bele Erembors', or 'Aiglentine', or 'Yolanz', or 'Amelot', or the like, who is sitting in her window and, more often than not, sewing. So, too, the Kölbigk lines have their *Merswyndam formosam*, and in the Scottish ballads the 'Bele' gets its echo in 'Fair', the *epitheton constans* for many an Annie, Eleanor, Isabel, Janet, Marjorie, or the like, who at the outset of a story is often similarly in the window of her bower and, it may be, 'sewing at her silken seam'.

The note, I trowe, imaked was in Fraunce.

MALORY

I<small>T</small> was, perhaps, his nostalgia for a decayed chivalry which led
William Caxton to make his greatest gift to English letters, the
so-called *Morte Darthur* of Sir Thomas Malory. The printing of
this was completed, about a year after the *Order of Chyualry* itself,
on 31 July 1485. Only a single perfect copy of it survives, which
is now in the Pierpont Morgan Library. Later editions were
printed by Wynkyn de Worde in 1498 and 1529 and William
Copland in 1557. Caxton's print has no initial title. A colophon describes it as 'thys noble and Joyous book entytled le
morte Darthur', and adds that it was 'reduced in to englysshe
by syr Thomas Malory knyght' and divided into books and
chapters by Caxton himself. In a prologue Caxton records the
request of 'many noble and dyvers gentylmen of thys royame of
Englond', which induced him 'to enprynte a book of the noble
hystoryes of the sayd kynge Arthur and of certeyn of his knyghtes
after a copye vnto me delyuered, whyche copye Syr Thomas
Malorye dyd take out of certeyn bookes of frensshe and reduced
it in to Englysshe'. And in the prologue to *Charles the Grete*,
which followed shortly after, he again refers to it as 'the book of
the noble and vyctoryous Kyng Arthur'. The term 'reduced',
in contemporary English, may signify either 'abbreviated' or
'translated', but it is clear, from other prologues by Caxton,
that it was in the latter sense that he employed it.

We do not know whether the 'copye' of Malory's romance
which Caxton obtained was in the hand of the author or in
that of a scribe. He is not likely to have been meticulous in his
adherence to it. His division into books and chapters often
breaks the thread of a continuous episode of narrative. And
certainly Malory himself would not have been guilty of the
lapse of French grammar involved in the *le Morte Darthur* of
the colophon. It is fortunate, with all respect to Caxton, that
we are no longer wholly dependent upon him for a text. In
1934 a manuscript of the romance was unexpectedly discovered
in the library of Winchester College. Preliminary studies of it
have been published by Mr. W. F. Oakeshott and Professor
Eugène Vinaver, and an edition by Professor Vinaver is in

preparation. It is certainly not Malory's own manuscript. This is clear from a passage referring to one following 'on the other side', which in fact begins on the same side in the manuscript. But it is probably of early date. A watermark on one page is almost identical with one on a document of 1475. The writing may well be of that date. And a fragment of vellum, used to mend a tear, comes from an indulgence of 1489 printed by Caxton himself. The text, however, differs so much from Caxton's version that it can hardly have been the one which he used. Unfortunately, a gathering is missing at each end, and we have, therefore, no comprehensive title. There are, however, several colophons, which come at the ends of some of the 'Tales' into which the romance is divided, and throw a little light upon the author as well as upon the structure of his work. The 'Tales' so treated are eight in number. The colophons, as a rule, provide some kind of a title for each. They are important, both critically and biographically, and must be given here. The first runs:

Here endyth this tale as the Freynshe booke seyth fro the maryage of kynge Uther unto kynge Arthure that regned aftir hym and ded many batayles. And this booke endyth whereas sir Launcelot and sir Trystrams com to courte. Who that woll make ony more lette hym seke other bookis of kynge Arthure or of sir Launcelot or sir Trystrams; for this was drawyn by a knyht presoner sir Thomas Malleorre, that God sende hym good recovery. Amen etc.
Explicit.

And the second:

Here endyth the tale of the noble kynge Arthure that was Emperoure hymself thorow dygnyté of his hondys. And here followyth, afftyr many noble talys, of sir Launcelot de Lake.

Explycit the noble tale betwyxt kynge Arthure and Lucius the Emperour of Rome.

The third:

Explicit a noble tale of sir Launcelot du Lake.

Here folowyth sir Garethis tale of Orkeney that was callyd Bewmaynes by sir Kay.

The fourth:

And I pray you all that redyth this tale to pray for hym that this wrote, that God sende hym good delyveraunce sone hastely. Amen.

Here endyth the tale of sir Gareth of Orkeney.

Here begynnyth the fyrste boke of syr Trystrams de Lyones, and

who was his fadir and hys modyr, and how he was borne and
fostyrd, and how he was made knyght of kynge Marke of Cornuayle.

The fifth:

Here endyth the secunde boke off syr Trystram de Lyones,
whyche drawyn was oute of freynshe by sir Thomas Malleorre,
knyght, as Jesu be hys helpe. Amen.

But here ys no rehersall of the third booke.

But here folowyth the noble tale off the Sankegreall, whyche
called ys the Holy Vessell and the Sygnyfycacion of blyssed bloode
off oure Lorde Jesu Cryste, whyche was brought into thys londe
by Joseph of Aramathye.

Therefore on all synfull, blyssed Lorde, have on thy knyght
mercy. Amen.

The sixth:

Thus endith the tale of the Sankgreal that was breffly drawy⟨n⟩
oute of Freynshe—which ys a tale cronycled for one of the trewyst
and of the holyest that ys in thys worlde—by sir Thomas Maleorre,
knyght.

O, blessed Jesu, helpe hym thorow Hys myght!
Amen.

The seventh:

And bycause I have loste the very mater of Chyvalere de Charyot
I departe from the tale of Sir Launcelot; and here I go unto the
Morte Arthur, and that caused Sir Aggravayne.

And here on the othir syde folowyth the moste pyteous tale of
the Morte Arthure Saunz Gwerdon par le Shyvalere Sir Thomas
Malleorre Knyght.

Jesu ayede ly pur voutre bone mercy! Amen.

For the final colophon, owing to the mutilation of the Winchester MS., we are dependent on Caxton, who prefixes it to
his own:

Here is the ende of the hole book of Kyng Arthur and his Noble
Knyhtes of the Rounde Table . . . And here is the ende of The
Deth of Arthur.

I praye you all, Jentylmen and Jentylwymmen that redeth this
book of Arthur and his Knyghtes from the begynnyng to the endynge,
praye for me whyle I am on lyve that God sende me good delyveraunce. And whan I am dede, I praye you all praye for my soule.

For this book was ended the ninth yere of the reygne of King
Edward the Fourth, by Syr Thomas Maleore, Knyght, as Jesu
helpe hym for hys grete myght, as he is the servaunt of Jesu bothe
day and night.

There is one other personal reference to the author which is worth noting. A commendation of Tristram for his learning in venery calls upon all gentlemen to praise him and pray for his soul. And here are added the words 'Amen, sayde Sir Thomas Malleore'.

It seems clear that the ungrammatical *le morte Darthur* of Caxton's colophon is merely a translation of *The Deth of Arthur*, which properly belongs to the eighth tale, and that Malory's own title for the romance as a whole was *The Book of King Arthur and his noble Knights of the Round Table*. For brevity we may perhaps refer to the individual tales as the *Coming of Arthur*, the *War with Rome*, the *Lancelot*, the *Gareth*, the *Tristram*, the *Sangreal*, the *Knight of the Cart*, which is again on Lancelot, and the *Death of Arthur*. It must be added that Malory's use of the term 'Tale' is not free from ambiguity. It does not always, especially in the earlier part of his story, indicate one of its main divisions. Sometimes there is a tale within a tale. Thus in the *Coming of Arthur* we find a reference to the 'book' of Balyn the Saveage, which later becomes a 'tale of Balyn and Balan' and has an 'Explicit' of its own. So, too, we get an 'Explicit the Wedding of Arthur' at the end of a long passage, which indeed begins with a mention of the wedding, but for the rest deals wholly with sporadic adventures of Sir Gawaine, Sir Torre, and King Pellinore. Many of the tales, indeed, are largely made up of strings of independent episodes, with abrupt transitions between them, which are indicated by recurrent phrases, on the models of the 'Or dit le conte', or more fully, for example, 'Ore laisse li contes à parler du chevalier de la charrete et retourne à parler d'une aultre matière', which are so common in French romance. Such are:

So leve we sir Tristram and turne we unto Kynge Marke

Now leve we of sir Lamorak and speke we of sir Gawayne

Here levith the tale of sir Launcelot and begynnyth of sir Percývale de Galis

Now turnyth thys tale unto sir Bors de Ganys

or, more unusually,

Here this tale overlepyth a whyle unto sir Launcelot.

Alternatively, the beginning of an episode is often indicated by a title for it in a marginal side-note.

It is possible that the opening leaves of the Winchester MS., now lost, may have contained some account of the sources from which Malory derived his material. His colophons to the *Coming of Arthur*, the *Tristram*, the *Sangreal*, and the *Knight of the Cart* make it clear that the chief of these was a 'Freynshe booke', and to this, or more briefly to 'the booke', there are further references in the text of all the tales, except the *War with Rome*. Professor Vinaver has made an elaborate investigation of the extant French Arthurian texts which cover Malory's ground and has come to the conclusion that, while his immediate exemplar cannot be precisely identified, it was probably one of a number of late compilations which were current in France during the fifteenth century, and were derivatives from an original prose cycle as it had developed in the thirteenth. It had a *Merlin*, a *Suite de Merlin* or *Livre d'Artus*, which gave the early history of the hero, a *Lancelot* or parts of one, possibly a *Gareih*, a *Tristan*, a *Queste del Saint Graal*, a *Mort Artu*, into which other parts of the *Lancelot* had been incorporated. Behind the prose cycle itself of course lay much earlier work, Geoffrey of Monmouth's quasihistorical narrative of the British champion, the Round Table contributed by Master Wace, the poems of Chrestien de Troyes, which brought in the Provençal motive of *amour courtois*, those of Robert de Boron, now mostly lost, the elaboration of the Grail story by the religious mysticism of the Cistercian writers. But Malory's French book was not his only source. He certainly also knew English books on his hero, and also traditions of him surviving in the countryside. In describing the departure of Arthur in a ship with three queens, he says 'I fynde no more wrytten in bokis that bene auctorysed', and he adds:

Yet som men say in many partyes of Inglonde that kynge Arthur ys nat dede, but had by the wyll of Oure Lorde Jesu into another place; and men say that he shall com agayne, and he shall wynne the Holy Crosse.

So, too, in describing the fate of Lancelot's kin, he adds:

And somme Englysshe bookes maken mencyon that they wente never oute of England after the deth of syr Launcelot—but that was but favour of makers! For the Frensshe book maketh mencyon—and is auctorysed, that syr Bors, syr Ector, syr Blamour and syr Bleoberis wente into the Holy Lande.

One English book, of which Malory made use, we are able to infer. It has long been thought that his account of the *War with Rome* was taken from a fourteenth-century alliterative *Morte Arthure*, of which a version is preserved in the Thornton MS. This he refers to as 'the Romaunce'. Traces of alliterative diction are apparent in Caxton's text. But it is now clear from the much fuller Winchester MS. that here Caxton has substituted a paraphrase of his own. The Winchester text is much longer and abounds in alliterative passages, evidently taken straight from the source. Malory, however, used a different version from that of the Thornton MS., and towards the end of the tale he abandoned it, since it contained an account of the rising of Mordred against Arthur, which was not to his purpose. It has been suggested that he also used for the *Death of Arthur* an English stanzaic *Morte Arthur*, now Harleian MS. 2252. Here, however, he does cite the French book, and it is more likely that both he and the writer of the English poem drew upon a French original no longer known. The one tale of which no other version, English or French, has been discovered is the *Gareth*. It has been suggested that this is of Malory's own composition, that the name Beaumains, which Gareth assumes, was adapted from that of Richard Beauchamp, Earl of Warwick, and that Malory drew upon a story told by the earl's biographer, John Rous, of a tournament at Calais, in which the earl appeared on three days in three different accoutrements, of which one was that of a Green Knight, and defeated all his opponents. This seems rather far-fetched. In fact Gareth makes no change of costume, although he does, but not in a tournament, successively slay or overthrow a Black, a Green, a Red, and a Blue Knight. Nor am I much impressed by the suggestion that the name of the Duke de la Rowse, who appears in a later episode, was taken from Rous himself. In fact, moreover, here as elsewhere, Malory cites the French book as his source. We can hardly, therefore, regard the request of the colophon that the readers will 'pray for hym that this wrote' as indicating an original composition.

It is possible that Malory began his work by writing the *War with Rome*, and then turned to the French book. There is a slight duplication of the embassy from the Emperor in the *Coming of Arthur*. And a passage in the colophon to this tale may conceivably suggest that at one time he meant to go no further:

Who that woll make ony more lette hym seke other bookis of kynge Arthur or of sir Launcelot or sir Trystrams.

But the eight tales, as they stand, are clearly meant to be read
as the successive chapters of a continuous narrative. There are
many references from tale to tale, and to some extent threads
dropped in one are taken up in another. What, then, must we
regard as the dominating theme which Malory was attempting
to develop? It sometimes looks as though there were none at
all. The French book was a dangerous model. Of some early
mythical elements in its origins little is left beyond Merlin and
the sorceries of the bad Morgan le Fay and the good Lady of the
Lake. When he reads that every day in the year, from early
morning to high noon, Gawaine's might increased unto thrice
its strength, the anthropologist may no doubt recognize that
originally Gawaine was a sun-hero. But, long before it came to
Malory, the narrative had been much elaborated by a succession of *remanieurs*, who had brought in innumerable sporadic
adventures on a common model. A knight rides abroad, meets
another, overthrows him, and sets free a lady whom he has
abducted, and rides away. A Gareth, a Palomides, a La Cote
Male Taille, conceived on these lines, easily becomes tedious.
Malory cut this element in his sources very freely, but not freely
enough. Here are still to the full the *Arturi regis ambages pulcher-
rimae*, of which Dante writes. Much of the *Tristram*, in particular,
which occupies over a third of Malory's pages, is largely irrele-
vant, and after all the tale, as he tells it, lacks its fine ending of
the black and white sails and the deaths of the lovers. Malory
begins well with the *Coming of Arthur*, but thereafter are many
confusions and inconsistencies. Merlin's prophecies are by no
means all confirmed. The origin of Arthur's sword Excalibur
is left very obscure. Galahad gets the description of the *haut
prince*, which properly belongs to a distinct Sir Galahalt.
Bagdemagus dies and is alive again later. There is a 'questing
beast', which is pursued by King Pellinor and later by Palomides, but we never learn its nature. There are many references
to the murders of Pellinor and his son Lamorak by Gawaine,
but the stories of them are never fully told. These are samples
only, and some of them may be due to Malory's sources rather
than himself. More important is the rift in the characterization
of Gawaine. In the *Coming of Arthur* Merlin speaks of him as the
man in the world whom Lancelot loves best, and anticipates
that Lancelot will slay him with Balin's sword. Lancelot does
not, in fact, although Galahad once defeats Gawaine with it.

Gawaine was held to be a good knight when he became one of the Round Table, and he did good service in Arthur's wars at home and against the Romans. But already he had slain a lady, although indeed it was by accident, and had been sworn by a quest of ladies to be courteous and always to show mercy. And in the tales that follow, his character is consistently depreciated. The deaths of Pellinor and Lamorak were to his discredit. Even his brother Gareth withdrew from his fellowship, 'for he was evir vengeable, and where he hated he wolde be avenged with murther: and that hated sir Gareth'. Tristram, too, describes his whole family, other than Gareth himself, as 'the grettyste distroyers and murtherars of good knyghtes that is now in the realme of Ingelonde'. It is rather surprising that he is the first knight to swear himself to the quest of the Grail. Galahad, however, refuses to ride with him. When bidden by a hermit, he will do no penance, and when a second hermit also reproves him he rides down the hill and abandons the adventure. But when we come to the *Death of Arthur*, this past reputation seems to have been largely forgotten. He is hostile to Lancelot because he had slain Gareth through an accident. But there is much nobility in the relations between them. He dies of a wound from Lancelot's hand, but writes him a letter, referring to 'all the love that ever was betwyxte us', and Lancelot in his turn weeps by his tomb and describes him as 'a full noble knyght as ever was born'.

The accounts of the Sangreal, again, are full of obscurities and even discrepancies. It has come into the narrative long before the tale devoted to the *Quest* of it. It is kept at Corbyn, which later becomes Carbonek, by a King Pellam, who later becomes Pelles. It contains part of the blood of Jesus Christ, which Joseph of Arimathea brought into the land. In the *Coming of Arthur* Balin comes to Pellam's castle, fights him, and wounds him with a marvellous spear which he found on a table. It was that of Longinus. Malory here refers to the *Quest*, and tells us that Pellam lay many years sore wounded, and might not be whole until Galahad healed him. It was 'the dolorouse stroke'. But when Lancelot visits Corbyn and begets Galahad on Elaine, the daughter of Pelles, and again when he runs mad and is harboured there as a fool, nothing is said of any infirmity. Pelles moves about, like any other king. Galahad, however, when he comes to Camelot, describes him as not yet whole

from Balin's stroke. Later he is more than once referred to as
the maimed king, but an entirely different account of the
maiming is given by Percival's sister. According to her Pelles,
while hunting in Ireland, came to a ship and found there
a sword, which he drew, and therewith entered a spear, with
which he was smitten through both thighs, and never since
might be healed, until Galahad came. It is probable that
Malory's French book had drawn upon more than one version
of the Grail story. And of this there is further evidence in a
greeting sent by Galahad from Camelot to 'my graunte-syre
kynge Pelles and unto my lorde kynge Pecchere', and in a
later passage where Lancelot is said to have seen the Grail in
'kynge Pescheors house'. Malory is not very lucid, again, in
distinguishing between Lancelot's imperfect vision of the Grail
and the more complete one vouchsafed to Galahad. Both of
them see much more than the covered vessel. Of Lancelot,
standing outside the door of a chamber, we are told,

> Before the Holy Vessell he saw a good man clothed as a pryste,
> and hit semed that he was at the sakeringe of the masse. And hit
> semed to sir Launcelot that above the prystis hondis were three men,
> whereof the two put the yongyste by lyknes betwene the prystes
> hondis, and so he lyffte hym up ryght hyghe, and hit semed to shew
> so to the peple.

The priest looked as if he were overcharged with the weight,
but when Lancelot attempted to enter the chamber and help
him, a breath of fire smote him, and he fell to the ground
insensible.

> 'Sir,' seyde they, 'the queste of the Sankgreall ys encheved now
> ryght in you, and never shall ye se of Sankgreall more than ye
> have seen.'

Galahad's experience is a much longer one. Joseph, the first
bishop of Christendom, comes in.

> And than the bysshop made sembelaunte as thoughe he wolde
> have gone to the sakeryng of a masse, and than he toke an obley
> which was made in lyknesse of brede. And at the lyfftyng up there
> cam a figure in lyknesse of a chylde, and the vysage was as rede
> and as bryght as ony fyre, and smote hymselff into the brede, that
> all they saw hit that the brede was fourmed of a fleyshely man.
> And than he put hit into the Holy Vessell agayne, and than he ded
> that longed to a preste to do masse.

The bishop vanished, and out of the Holy Vessell came a man
'that had all the sygnes of the Passion of Jesu Cryste, bledynge
all opynly'. He took the Vessell, and from his hands Galahad
and his fellows received their Saviour. This experience goes
beyond Lancelot's, to whom no word was said by the Divine
Person, but both alike have entailed what a theologian would
call the Real Presence.

The entanglements are over when the Grail has vanished,
and thereafter Malory is free to tell one of the best stories of
the world. Lancelot resumes his love of Guenevere, forgetting
'the promyse and the perfeccion that he made in the queste'.
Aggravayne spreads the scandal. Lancelot withdraws from
Guenevere, to her anger, and leaves the court. Guenevere is
falsely accused of poisoning. She must be burnt, unless a knight
will do battle for her. Lancelot does, and the lovers are reconciled. Then comes the story of the love of Elaine of Astolat for
Lancelot, an episode again, but a gracious one. Malory now
allows himself one of his rare comments, in a discourse on May
time, and the instability of modern love, as compared with that
of King Arthur's days. It was not so with Queen Guenevere,
'for whom I make here a lytyll mencion, that whyle she lyved
she was a trew lover, and therefor she had a good ende'. She
goes maying, is abducted by Mellyagaunce, and rescued by
Lancelot, who drops his dignity to ride in a cart. Both king and
queen now cherish Lancelot. Another episode comes to exalt
him, in his healing of the wounds of Sir Urre, after which he
weeps 'as he had bene a chylde that had been beatyn'. But now
tragedy is at hand. It is introduced by an ironical repetition of
the May-day theme. Aggravayne is still plotting, and with
him his half-brother Mordred, Arthur's illegitimate son. They
denounce the amour of Lancelot and Guenevere to Arthur, who
indeed already had 'a demyng of hit'. Then they have the luck
to find Lancelot in Guenevere's chamber. Lancelot slays
Aggravayne and wounds Mordred. Mordred reports the affair
to Arthur, who condemns Guenevere to be burnt at the stake.
Gawaine refuses to be present, and his brothers Gaherys and
Gareth will only go unarmed. Lancelot comes to rescue the
queen, and takes her to Joyous Garde. But in the mellay he
has the misfortune to slay Gaherys and Gareth. As a result,
Gawaine becomes his relentless foe. He instigates Arthur to
besiege Joyous Garde. Lancelot, from its walls, asserts the

innocence of Guenevere and upbraids Gawaine with the death
of Lamorak. Gawaine will not allow Arthur to be reconciled,
and a battle follows. Lancelot bids his knights to spare Arthur
and Gawaine, and when Arthur is unhorsed, alights to remount
him, and Arthur weeps, 'thynkyng of the grete curtesy that was
in sir Launcelot more than in ony other man'. Then the Pope
intervenes, bidding Arthur take Guenevere again, and accord
with Lancelot. Arthur is appeased, but not Gawaine, although
Lancelot offers to undertake a pilgrimage for the souls of his
brothers. Lancelot now goes overseas to his realm of Benwick.
He suspects that Mordred will make trouble in England.
Arthur, instigated by Gawaine, takes an expedition against
Benwick, leaving Mordred in charge at home. Gawaine and
Lancelot again dispute and fight. Gawaine is wounded, but
Lancelot spares him. In a second fight he is smitten by Lancelot
on the same wound. Now comes news that Mordred has
declared himself king. He would have wedded Guenevere, but
she takes refuge in the Tower of London. Arthur returns to
England and defeats him. Gawaine is again smitten on his old
wound. He writes to Lancelot, bidding him visit his tomb and
pray for his soul, and 'for all the love that ever was betwyxte us'
rescue Arthur from Mordred. And he bids Arthur cherish
Lancelot. Arthur has already defeated Mordred at Dover and
Barham Down. But Mordred is strong in the counties around
London. The ghost of Gawaine advises Arthur in a dream to
put off battle until Lancelot comes. There is a parley between
the king and his illegitimate son. But the accidental drawing of
a sword to kill an adder provokes a fresh battle in which
Mordred falls, and Arthur is stricken to the death. He bids
Bedivere throw his sword Excalibur into a water, and a queen,
with three ladies in black hoods, arrives and takes him in a
barge to be healed of his wound in the Isle of Avilion. Guenevere becomes a nun at Amesbury. Here Lancelot finds her
when he returns to England. They part in piety. Lancelot
becomes a priest. He visits Amesbury once more to see Guenevere dead. He buries her at Glastonbury, dies himself, and is
taken to Joyous Garde, where Ector speaks the famous eulogy
over his corpse. Constantine reigns in England.

What, then, was the dominant motive which we may suppose
to have inspired Malory in making his careful selection from the
very amorphous material of his French book? Professor Vinaver

has suggested that he was a practical and righteous fifteenth-century gentleman, who wished to bring back a decadent England to the virtues of 'manhode, curtesye and gentylnesse', which he believed to have inspired the 'custom and usage' of medieval chivalry. This would certainly have appealed to Caxton. Obviously chivalry looms large in the romance. The Knights of the Round Table are sworn:

Never to do outerage nothir morthir, and allwayes to fle treson, and to gyff mercy unto hym that askith mercy, uppon payne of forfiture othir worship and lordship of kynge Arthure for evirmore; and allwayes to do ladyes, damesels, and jantilwomen and wydowes strengthe hem in hir ryghtes and never to enforce them uppon payne of dethe. Also that no man take no batayles in a wrongefull quarell for no love ne for no worldis goodis.

And chivalrous adventure is the obligation of noble birth. The mother of Aglovale and Percyvale would have them abide at home.

'A, my swete modir,' seyde sir Percyvale, 'we may nat, for we be comyn of kynges bloode of bothe partis. And therefore, modir, hit ys oure kynde to haunte armys and noble dedys.'

It is a very aristocratic ideal. Nobody counts for much, except the knights and their ladies, and a sprinkling of pious hermits. And even these are of aristocratic origin.

For in thos dayes hit was nat the gyse as ys nowadayes; for there were none ermytis in tho dayes but that they had bene men of worship and of preuesse, and tho ermytes hylde grete householdis and refreysshed people that were in distresse.

The rest are 'churls' and of no account. They take to their heels when they see a knight coming. And no wonder! A carter, busy on fetching wood for his lord, refused to stop his work and give Lancelot a ride to a castle. And then

Sir Launcelot lepe to hym and gaff hym backwarde with hys gauntelet a reremayne, that he felle to the erthe starke dede.

I doubt whether Malory has his eye much on the England of his own day. He does, indeed, stop, in the middle of his account of the rising of Mordred against Arthur, to make one of his rare personal comments:

Lo ye, all Englysshemen, se ye nat what a myschyff here was? For he that was the moste kynge and nobelyst knyght of the worlde, and most loved the felyshyp of noble knyghtes, and by hym they all were upholdyn, and yet myght nat these Englysshemen holde them contente with hym. Lo thus was the olde custom and usayges of thys londe, and men say that we of thys londe have nat yet loste that custom. Alas! thys ys a greate defaughte of us Englysshemen, for there may nothynge us please no terme.

But of the England of the fifteenth century, exhausted by generations of foreign enterprise and dynastic quarrels, of England as we find it depicted in the *Paston Letters*, of the complete breakdown of law and order, of the abuses of maintenance and livery and private warfare, of the corruption of officials, of the excessive taxation, of the ruin of countrysides by the enclosure of agricultural land for pasture—of all this we find no consciousness whatever in Malory's pages. A revival of the spirit of chivalry might have done something to help matters, but a strong hand in the central government would have done more.

Malory does not, however, except in this outburst, come before us as a political thinker, but as a story-teller, intent on the development of a very dramatic theme. It is the drama of Arthur and Lancelot, and indeed, as it works out, it seems to be Lancelot, rather than Arthur, who is the protagonist. Arthur is prominent at the beginning, with his triumphant coming, the success of his early wars, the mystery of Excalibur, and the establishment of the Round Table. Later his rôle becomes a more passive one. His begetting of Mordred is significant. Between Arthur and Lancelot stand Guenevere and later Gawaine. To Arthur belong Merlin, Nimue, and Morgan le Fay. Merlin drops early out of the story, but he lives long enough to prophesy of Lancelot, and to see Lancelot himself as a boy. And among all the entanglement of Malory's earlier tales, the most notable thing is the constant emphasis on Lancelot. In the *War with Rome* there are frequent mentions of him, which appear to be Malory's own additions to the alliterative poem as he found it. Certainly they are not in the version of the Thornton MS. It is difficult to be sure, without greater knowledge of Malory's French book than we possess, but it seems likely enough that the ruthless cuttings of it, which we conjecture, were largely motived by the desire to bring Lancelot into the foreground. The third tale is entirely devoted to him, and at the end of it his pre-eminence among the members of the Round Table is assured.

At that tyme sir Launcelot had the grettyste name of ony knyht of the worlde, and moste he was honoured of hyghe and lowe.

Thereafter he is constantly referred to as the exemplar of perfect knighthood. In the long-drawn-out earlier part of the *Tristram* the only significant thing is the friendship which establishes itself between him and Tristram. 'Of all knyghtes', says Tristram, 'he bearyth the floure.' It is noteworthy, again, that Malory brings Lancelot as near to the complete vision of the Grail as he can, and that of those who ultimately achieve it, one is his son and another his nephew. Of his part in the final outcome of the drama no more need be said. But it is relevant that it is on him, and not on Arthur, that the threnody, which forms its epilogue, is spoken.

If we feel that Malory is slow in getting under way with his high theme and too often allows the earlier part of his narrative to be unduly clogged with episodes, there can at least be no doubt as to the singular beauty of his prose. He is free from the desire to 'augment' the English language with 'aureate' terms, which is the bane of so much contemporary writing, and comes nearer to the vernacular tradition which Dr. R. W. Chambers has traced back, through the pieties of the thirteenth and fourteenth centuries, to its Anglo-Saxon beginnings. But he gives it a wider scope by applying it to secular material, and here, of course, he is much under the influence of his French models. Professor W. P. Ker, who has said so many of the best things about medieval literature, speaks of his 'high imaginative prose', and in elaboration:

The style of his original has the graces of early art; the pathos, the simplicity of the early French prose at its best, and always that haunting elegiac tone or undertone which never fails in romance or homily to bring its sad suggestions of the vanity and transience of all things, of the passing away of pomp and splendour, of the falls of princes. In Malory, while this tone is kept, there is a more decided and more artistic command of rhythm than in the Lancelot or the Tristan. They are even throughout, one page very much like another in general character: Malory has splendid passages to which he rises, and from which he falls back into the even tenour of his discourse. In the less distinguished parts of his book, besides, there cannot fail to be noted a more careful choice of words and testing of sounds than in the uncalculating spontaneous eloquence of his original.

One may add that Malory, unlike Caxton, does not share the love of his French predecessors for linking synonymous words in doublets, and that he makes no attempt to reproduce in English those elaborate periods of carefully linked subordinate clauses which they favoured. He prefers to proceed, both in narrative and in dialogue, by a succession of simple sentences, each introduced with an 'And', 'But', 'So', 'Then', 'For', 'Wherefore', or the like, and to obtain his rhythm by balancing these with longer ones. He has been accused of occasional breakdowns in grammatical construction, but a comparison of the Winchester text with that of Caxton suggests that here the scribes have not always served him well. The diction is fairly modern. There are a few French words which have not established themselves in the language, and a few English ones which have died out of it. Dialogue is apt to come to an abrupt conclusion. The knights are men of their hands and have no turn for a prolonged debate. Arthur calls them together to plan a coming campaign. 'They coude no counceil gyve, but said they were bygge ynough.' Lancelot is advised to avoid a wrong quarrel. 'As for that,' said Sir Lancelot, 'God ys to be drad.' But a hermit can outdo them in brevities of speech. One has called upon Sir Gawaine to do penance for his sins, but Sir Gawaine refuses. ' "Well", seyde the good man, and than he hylde hys pece.' Malory is fond of vivid words. Knights come into battle 'as hit had bene thunder', or 'hurtling', or with a 'wallop'. They fall to the earth 'flatling' or 'noseling'. A spear comes 'poyntelynge'. There is little deliberate scenic description, but a strong feeling for out-of-door life. An army stands still 'as hit had be a plumpe of woode'. A knight lashes at a shield, 'that all the medow range of the dyntys'. He rides by moonlight, and comes to 'a rowghe watir, which rored', or finds 'an ermytage, whiche was undir a woode, and a grete cliff on the othir syde, and a fayre watir rennynge undir hit'. It has the simplicity of an Italian miniature. And in passages of spiritual exaltation Malory can rise without effort to the level of his theme. Galahad comes to court, and Lancelot looks upon his son, as he knights him. 'God make you a good man, for of beauté fayleth you none as ony that ys now lyvynge.'

Who then was Sir Thomas Malory? Professor Rhŷs, an ardent Celticist, has adopted the early statement of John Bale that he was a Welshman, and indeed he has introduced into his *Tristram* several Welsh knights, whom he did not find in his French source. But the name Malory, variously spelt, is traceable in Yorkshire, Northamptonshire, Leicestershire, Cambridgeshire, and Warwickshire. Only one Sir Thomas has, however, been identified, and to him most recent scholars have been inclined to ascribe the romance. He came of a Northamptonshire family, originally settled at Draughton and later at Winwick, and at Swinford in Leicestershire. A marriage of Sir Stephen Malory in the fourteenth century to Margaret Revell of Newbold Revel, also called Fenny Newbold, in Warwickshire, brought that estate into the family. Their grandson, John Malory, was born before 1383, held various offices in Warwickshire, was M.P. for the county in 1413, and died about 1434, and his son was Sir Thomas, who was in his turn M.P. in 1444 or 1445, and died in 1471, leaving a widow Elizabeth. Dugdale tells us that 'in K.H.5 time' he 'was of the retinue to Ric. Beauchamp E. Warr. at the siege of Caleys, and served there with one lance and two archers'. He was buried in the chapel of St. Francis at the Grey Friars in London, where the inscription on his tomb runs 'dñs Thomas Mallere, valens miles, ob. 14 Mar. 1470, de parochia de Monkenkyrkby in comitatu Warwici'. In modern dating '1470' should, of course, be '1471'. Contemporary records describe him sometimes as of Newbold and sometimes as of Monks Kirby. The places are close together. There is an obvious error in Dugdale's account, since there was no siege of Calais in the reign of Henry V. But it is not so obvious what it was. Possibly he wrote 'Caleys' for 'Harfleur' (1415) or Rouen (1418). But an easier slip would be to write 'K.H.5' for 'K.H.6', and there was in fact a siege of Calais in 1436, at which Beauchamp was present. If Malory was of age to hold a lance at Harfleur or even Rouen, he would have been an old man at the time of his death. Modern research has revealed a good deal about him from 1443 onwards, and it is disconcerting. In 1443 Thomas Smythe of Sprotton, Northants., charged him and another with assault and the taking of goods and chattels to the value of £40. The sheriff was ordered to attach him, and did so, but no more is known of the case. Equally inconclusive is the record of a complaint in Chancery, which may be of a date during 1443–50, by Katherine, Lady Peyto, that he had entered with armed men into her manor of Syburtoft, Northants., and driven off four 'rotherbestes', or cattle, owned by her bailiff to his home in Warwickshire. These and some others of the exploits ascribed to Malory may be merely examples of the high-handed methods of recovering debts which arose from the break-down of legal process in the fifteenth century. But from 1451 onwards the record becomes clearer. On 13 July of that year a royal warrant directed Humphrey Stafford, Duke of Buckingham, and Richard Neville, Earl of Warwick, to arrest Malory and his servant John Appleby, and bind them under bail to do no hurt to the Carthusian house of Axholme or any of the King's people, and to appear before the King and Council after Michaelmas to answer certain charges. This clearly relates to the priory of Monks Kirby, once a cell of the monastery of St. Nicholas of Angiers, which had been transferred to that of Epworth in Axholme, Lincolnshire, on the suppression of alien priories. Malory was accordingly arrested on 25 July and committed to the Sheriff of Warwickshire at Coleshill. What followed we learn from an inquisition by a grand jury, sitting under the Duke of Buckingham and other Justices of the Peace at Nuneaton on 23 August. According to the findings of the jury, Malory, two days after his arrest, had escaped by swimming the Coleshill moat. On the two following days, in company with Appleby and others, he had raided the Cistercian Abbey of Combe, near Coventry, insulted the abbot and his monks, and carried off large sums of money, together with jewels and other valuables. The jury go on to record certain other acts of violence which had taken place before his arrest, during 1450. Some of these may again have been merely high-handed enforcements of debt claims. But two of them were of more importance. On 4 January he had collected an armed force of twenty-six men and lain in wait among the woods of Combe Abbey with the intent of murdering the Duke of Buckingham. It is to be hoped that the duke hitched back his chair while evidence was given as to this. On 25 May he had broken into the house of Hugh Smyth at Monks Kirby and raped his wife Joan, and on 6 August he had repeated this offence at Coventry, and had taken her, with goods worth £40, to Barwell in Leicestershire. Apologists for Malory have endeavoured to minimize these last charges. It is absurd, they say, to suppose that Malory would rape the same woman twice, and probably all that happened was that Joan was roughly handled while her husband's goods were rifled. This would be technically *raptus*. But the finding of the jury is precise in both cases—*cum ea carnaliter concubuit*. A second inquisition of the same

time and place, but with a different jury, records another raid
on cattle and sheep in June 1451 at Cosford in Warwickshire.
Further proceedings against Malory became matter for the
High Court. In Michaelmas 1451 Hugh Smyth brought an
action for *raptus* and breach of peace. The defendants did not
appear. An order to the sheriff to attach their goods was
answered by a statement, obviously untrue, that they had none.
An arrest for trial in Hilary term 1452 was then directed, and
of this case we hear no more. But during the same term a new
charge was brought against Malory by John, Duke of Norfolk,
and his wife Eleanor, John Stafford, Archbishop of Canterbury,
and Humphrey, Duke of Buckingham. He had broken on 20
July 1451 into their park of Caludon, near Coventry, taken
six does, and done other damage to the value of £500. A
warrant for his arrest and that of Appleby and his other
associates appears to have been directed by the High Court on
15 March 1452. One recorded in the Patent Rolls on 26 March,
and addressed to the Duke of Buckingham, Sir Edward Grey
of Groby, and the sheriff, is not specific as to 'certain charges',
which he is to answer. In any case, Malory made an appearance
on the Caludon charge and put himself upon his country.
Order was given for the empanelling of a jury. None was
forthcoming on the appointed day, and after postponements
during a year nothing more of the case is on record. The
failure, through official incompetence or corruption, to secure
juries is one of the worst features of fifteenth-century mis-
government. Precisely the same thing happened with regard
to the main charges against Malory, which arose out of the
Nuneaton inquisitions. John Appleby and his other con-
federates were outlawed at Warwick on 21 August 1452. But it
was not until January 1453 that Malory once more put himself
upon his country at Westminster. Repeated postponements of
his trial followed, during which he was variously held by the
Sheriffs of London at Newgate, by the King's Marshal at the
Marshalsea, or by the Constable of the Tower. In May 1454
he was temporarily released, on the bail of Roger Chamburleyn
and others, to 29 October. On that day he did not appear, and
his sureties reported that he was in jail at Colchester in Essex.
Of his activities in the interval we learn from inquisitions held
at Chelmsford and Braintree. He had been harbouring and
inciting to various felonies his servant John Addelsey and others,

had been committed to jail, and once more had broken out. By 18 November the London authorities had recovered him, and again, during a year's space, no jury was forthcoming. On 24 November 1455 he obtained royal letters patent for a pardon, which he produced in court a few months later. How far he profited by them remains rather obscure. He had creditors also to satisfy, and at one time he appears to have been in Ludgate, which was a civil prison for debtors. But for a short period in 1457 he was bailed out to William Neville, Lord of Fauconberg, a kinsman of the Earl of Warwick. For 1458 we have no record. In the spring of 1459 he was reported to be in Warwickshire when he ought to have been in the Marshalsea. Early in 1460 he was once more committed to Newgate. But presumably he was a free man and not in disgrace by December 1462, since his name then occurs in a list of *milites* who accompanied Edward IV on an expedition to the north. It has been suggested by Mr. Hicks that the 'Thomas Malery' of this list was another man, on the ground that 'Sir' is prefixed to the name of each individual knight in the list, and the 'Thomas Malery' here mentioned has no such prefix. But this is fantastic. The list, printed by James Gairdner in his *Three Fifteenth-Century Chronicles* (p. 157), is headed 'Thes be the namys of dewkes, erlys, barons, and knytes beyng with owre soveryn Lord Kyng Edward'. Then follow the names of two dukes, and then groups of names, headed respectively 'Erlys', 'Barons', and 'Milites'. The 'Milites' can only be the 'knytes' of the general heading. There are fifty-nine of them, including 'Thomas Malery'. It is not the case that 'Sir' is prefixed to the name of each knight. 'Ser' precedes the first five names and six of the others. Its omission elsewhere must be a mere scribal irregularity. I have stressed this, because the appearance of Sir Thomas Malory in Edward's train tends to invalidate the suggestion which has sometimes been made that he was a Lancastrian, and had perhaps fought on the Lancastrian side in the Wars of the Roses. This seems to me unlikely on other grounds. While a Warwickshire man would naturally follow Beauchamp, Earl of Warwick, in the time of Henry V, he would as naturally take the side of Neville, Earl of Warwick, during the dynastic struggle. It was a Neville of this house who stood surety for him in 1457. On the other hand, his chief local enemy was clearly Humphrey Stafford, Duke of Buckingham, who died fighting for the Lancastrian cause at Northampton in 1460. And it is worth noting that the pardon which Malory received on 24 November 1455, although in the form of a royal patent, must actually have been granted by the Duke of York, who was then acting as Protector during one of King Henry's disabling illnesses. Finally, if Malory had fought as a Lancastrian, his name would surely have appeared, as it does not, in the long lists of persons attainted, which are recorded in the Rolls of Parliament during 1461 and 1465. He was probably the Sir Thomas Malory who is noted by John Nichols, in his *History of Leicestershire*, as witnessing a conveyance for a member of the Feilding family in 1464, since the Feildings, later at any rate, were of Newnham Paddox in Warwickshire. We know nothing more of him until 1468, when he was excluded from the operation of two general pardons granted by Edward IV on 24 August and 1 December. The reason for this remains quite obscure. Possibly he was involved in the intrigues of the Earl of Warwick against the king, which began in 1466, but did not result in an open breach before 1469. There is nothing to show that he was in prison at the time. Mr. T. W. Williams has suggested that he may have gone to Bruges and may there have come into contact with Caxton and given him a copy of his romance. An amnesty granted by Edward at the end of 1469 may have enabled him to return to England. Alternatively, he may have committed some further offence against the rights of the Carthusians of Axholme in the priory of Monks Kirby, which received a new royal confirmation in 1468–9. We are reduced to guess-work. All that we really know is that he died on 14 March 1471, presumably in England, since he was buried in the church of the Grey Friars in London.

There are some *prima facie* reasons for accepting this Sir Thomas Malory of Newbold Revell and Monks Kirby as the author of the romance. One, which is negative, rather than positive, is that no other contemporary Thomas Malory, who was a knight, has been traced. A second is, of course, to be found in the repeated references in the colophons of the romance itself to the writer as a prisoner. They are sometimes accompanied or replaced by expressions of piety and contrition, although these are perhaps such as any Christian man might write. They go on to the end of the book, which suggests that its Sir Thomas was still or again a prisoner in 1469 or 1470.

And to them may be added an interesting passage in the *Tristram*, which is not taken from Malory's source:

So sir Trystram endured there grete payne, for syknes 'had undirtake hym, and that ys the grettist payne a presoner may have. For all the whyle a presonere may have hys helth of body, he may endure undir the mercy of God and in hope of good delyveraunce; but when syknes towchith a presoners body, than may a presonere say all welth ys hym berauffte, and than hath he cause to wayle and wepe.

It is noteworthy, again, that Sir Thomas of Monks Kirby was buried not at home but in the church of the Grey Friars, which is close to the prison of Newgate.

Incidentally, it would be easier to think of the romance-writer than of the Warwickshire knight as a Lancastrian. King Mark, faced with a revolt of his knights, vows to go to Rome and war against the miscreants, and adds,

I trow that is fayrer warre than thus to areyse people agaynste youre kynge.

This is not very conclusive, if it is meant as a political reference, since Englishmen were divided as to whether the dynastic right to rule lay with the house of Lancaster or with that of York. More significant, perhaps, is a passage on the treachery of Mordred:

Than Sir Mordred araysed muche people aboute London, for they of Kente, Southsex and Surrey, Esax, Suffolke and Northefolke, helde the moste party with Sir Mordred.

It was precisely in the districts here named that the Yorkist strength lay. However this may be, it is difficult to resist the feeling that there is a marked spiritual cleavage between the Malory of romance and the Malory whom recent biographical research has revealed.

'What?' seyde sir Launcelot, 'is he a theff and a knyht? and a ravyssher of women? He doth shame unto the Order of Knyghthode, and contrary unto his oth. Hit is pyté that he lyvyth'.

Surely the Sir Thomas of Monks Kirby could not have written this without a twinge.

BIBLIOGRAPHY[1]

GENERAL

A FUNDAMENTAL work is J. E. Wells, *A Manual of the Writings in Middle English, 1050–1400* (New Haven, 1916), with bibliographical notes and eight *Supplements* to 1942. A promised volume, covering the fifteenth century, has not yet appeared. Also of great value is the *Index of Middle English Verse* by Carleton Brown and R. H. Robbins (New York, 1943), which indexes all Middle English poems. F. W. Bateson, *The Cambridge Bibliography of English Literature* (1940) and L. L. Tucker and A. R. Benham, *A Bibliography of Fifteenth Century Literature* (Seattle, Washington, 1928) are useful. Current work is annually recorded in *The Year's Work in English Studies* (1919–20 onwards, English Association) and *The Annual Bibliography of English Language and Literature* (1920 onwards, Modern Humanities Research Association). Useful also are *The British Museum Catalogue of Printed Books* (new ed. in progress), *The Dictionary of National Biography*, and C. S. Northup, *Register of Bibliographies of the English Language and Literature* (Yale, 1925).

Collections and anthologies, covering more than one of the following chapters, are T. Wright and J. O. Halliwell, *Reliquiae Antiquae* (1841, 1845); W. C. Hazlitt, *Remains of the Early Popular Poetry of England* (1864–6); T. H. Ward and others, *The English Poets* (1887–94); D. Laing, *Select Remains of the Ancient Popular Poetry of Scotland* (1884); E. Flügel, *Neuenglisches Lesebuch* (Halle, 1895); A. W. Pollard, *Fifteenth Century Prose and Verse* (1903); F. Kluge, *Mittelenglisches Lesebuch* (Halle, 1904); O. F. Emerson, *Middle English Reader* (New York, 1905, 1915); A. S. Cook, *Literary Middle English Reader* (Boston, 1915); A. Brandl and O. Zippel, *Mittelenglische Sprach- und Literaturproben* (Berlin, 1917,

---

[1] Abbreviations are: *Archiv*=(Herrig's) *Archiv für das Studium der neueren Sprachen* (various places); EETS=Early English Text Society; *ELH*=*English Literary History* (Baltimore); *ES* = *Englische Studien* (Leipzig); *JEGP*=*Journal of English and Germanic Philology* (Bloomington, Indiana); *MLN*=*Modern Language Notes* (Baltimore); *MLQ* = *Modern Language Quarterly*; *MLR* = *Modern Language Review*; *MP*= *Modern Philology* (Chicago); *NQ*=*Notes and Queries*; *PMLA*=*Publications of the Modern Language Association of America* (New York); *PQ* =*Philological Quarterly* (Iowa City); *RES*=*Review of English Studies*; SATF=Société des Anciens Textes Français (Paris); *SP*=*Studies in Philology* (Chapel Hill); STS=Scottish Text Society; *TLS* = *Times Literary Supplement*.

1927); G. Sampson, *The Cambridge Book of Prose and Verse* (1924);
G. H. Gerould, *Old English and Medieval Literature* (New York,
1929).

Among relevant histories of English literature may be noted
T. Warton, *History of English Poetry from the Twelfth to the Sixteenth
Century* (1774–81, ed. W. C. Hazlitt 1871); H. Taine, *Histoire de
la littérature anglaise* (Paris, 1863–4, tr. 1871); B. ten Brink,
Geschichte der englischen Litteratur (Berlin, 1877–93, tr. 1883–96,
vol. i revised A. Brandl, 1899); G. Körting, *Grundriss der Geschichte
der englischen Litteratur* (Münster, 1887, 1910); H. Morley,
English Writers (1887–95); J. J. Jusserand, *Histoire littéraire du
peuple anglais des origines à la Renaissance* (Paris, 1894, tr. 1895);
W. J. Courthope, *History of English Poetry* (1895–1910); R.
Wülcker, *Geschichte der englischen Litteratur* (Leipzig, 1896, 1906–
7); G. Saintsbury, *A Short History of English Literature* (1898);
G. Gregory Smith, *The Transition Period* (1900); F. J. Snell, *The
Age of Transition* (1905); W. H. Schofield, *English Literature from
the Norman Conquest to Chaucer* (1906); A. W. Ward and A. R.
Waller, *The Cambridge History of English Literature* (1907–16,
1932); W. P. Ker, *English Literature, Medieval* (1912); J. M.
Berdan, *Early Tudor Poetry* (New York, 1920, 1931); E. Legouis
and L. Cazamian, *Histoire de la littérature anglaise* (Paris, 1924, tr.
1926–7); the introduction to E. P. Hammond's *English Verse
between Chaucer and Surrey* (N. Carolina, 1927); O. Elton, *The
English Muse* (1933); W. F. Schirmer, *Geschichte der englischen
Literatur von den Anfängen bis zur Gegenwart* (Halle, 1937). A
background is given by H. Paul, *Grundriss der germanischen Philo-
logie* (Strassburg, ed. 2, 1896–1903); G. Gröber, *Grundriss der
romanischen Philologie* (Strassburg, 1897–1906). On metric are
J. Schipper, *Englische Metrik in historischer und systematischer
Entwicklung* (Bonn, 1881–8) and *Grundriss der englischen Metrik*
(Vienna, 1895, tr. 1910); G. Saintsbury, *A History of English
Prosody* (1906–10); M. Kaluza, *Englische Metrik in historischer
Entwicklung dargestellt* (Berlin, 1909, tr. 1911); J. P. Oakden,
Alliterative Poetry in Middle English (1930–5).

CHAPTER I. MEDIEVAL DRAMA

The most comprehensive account of European medieval
drama in general is W. Creizenach, *Geschichte des neueren Dramas*
(Halle, ed. 2, 1911–23). It may be supplemented by W. Cloetta,
Beiträge zur Litteraturgeschichte des Mittelalters und der Renaissance

(Halle, 1890–2), and K. Mantzius, *Skuespilkunstens Historie* (Copenhagen, 1897–1907, tr. 1903–21); for France by L. Petit de Julleville, *Histoire du Théâtre en France* (Paris, 1880–6), and E. Lintilhac, *Histoire générale du Théâtre en France* (Paris, 1904–9).; for Italy by A. D'Ancona, *Origini del Teatro italiano* (Turin, 1891); and for England by J. P. Collier, *History of English Dramatic Poetry* (1831, 1879, an untrustworthy work); A. W. Ward, *A History of English Dramatic Literature to the Death of Queen Anne* (1875, 1899); J. J. Jusserand, *Le Théâtre en Angleterre* (Paris, ed. 2, 1881); J. A. Symonds, *Shakespeare's Predecessors in the English Drama* (1884, rev. ed., 1900); F. E. Schelling, *Elizabethan Drama* (Boston, 1908), and the present writer's *The Mediaeval Stage* (1903).

A full and admirable account of the liturgical plays and their relation to the liturgy itself, with a *corpus* of the more important texts, is given by K. Young, *The Drama of the Medieval Church* (1933).

Other studies and collections are F. J. Mone, *Schauspiele des Mittelalters* (Karlsruhe, 1846); E. du Méril, *Origines latines du Théâtre moderne* (Paris, 1849, 1897); C. A. von Hase, *Das geistliche Schauspiel des Mittelalters* (Leipzig, 1858, tr. 1880); C. E. de Coussemaker, *Drames liturgiques du Moyen Âge* (Rennes, 1860); E. Wilken, *Geschichte der geistlichen Spiele in Deutschland* (Göttingen, 1872); M. Sepet, *Le Drame chrétien au Moyen Âge* (Paris, 1878, tr. 1880) and *Origines catholiques du Théâtre Moderne* (Paris, 1901); G. Milchsack, *Die lateinischen Osterfeiern* (Wolfenbüttel, 1880); A. Reiners, *Die Tropen-, Prosen- und Präfations-Gesänge des feierlichen Hochamtes im Mittelalter* (Luxemburg, 1884); L. Gautier, *Histoire de la Poésie liturgique au Moyen Âge* (Paris, 1886); K. Lange, *Die lateinischen Osterfeiern* (Munich, 1887); R. Froning, *Das Drama des Mittelalters* (Stuttgart, 1891); A. Gasté, *Les Drames liturgiques de la Cathédrale de Rouen* (Évreux, 1893); R. Heinzel, *Beschreibung des geistlichen Schauspiels im deutschen Mittelalter* (Hamburg, 1898); G. Cohen, *Histoire de la Mise en Scène dans le Théâtre religieux français du Moyen Âge* (Paris, 1906, 1926) and *Le Théâtre religieux* (Paris, 1928); J. S. Tunison, *Dramatic Traditions of the Dark Ages* (Chicago, 1907); V. De Bartholomaeis, *Le Origini della Poesia drammatica italiana* (Bologna, 1924); A. Jeanroy, *Le Théâtre religieux en France du xie au xiiie siècles* (Paris, 1924); W. Stammler, *Das religiöse Drama im deutschen Mittelalter* (Leipzig 1925); E. A. Wright, *The Dissemination of the Liturgical Drama in*

France (Bryn Mawr, 1936). The *scholares vagantes* can be studied in J. J. Champollion-Figeac, *Hilarii Versus et Ludi* (Paris, 1838); J. A. Schmeller, *Carmina Burana* (Breslau, 1847, 1894); A. Hilka and O. Schumann, *Carmina Burana* (Heidelberg, 1930 sqq.); H. Waddell, *The Wandering Scholars* (1927); F. J. E. Raby, *A History of Christian Latin Poetry from the Beginnings to the Close of the Middle Ages* (1927) and *A History of Secular Latin Poetry in the Middle Ages* (1934). The range of the Easter plays may be inferred from H. J. Feasey, 'The Easter Sepulchre' (1905, *Ecclesiastical Review*, xxxii. 337–55, 468–99); N. C. Brooks, *The Sepulchre of Christ in Art and Liturgy* (1921, University of Illinois Studies in Language and Literature, vii. 2) and the same writer's 'The *Sepulchrum Christi* and its Ceremonies in Late Mediaeval and Modern Times' (1928, JEGP xxvii. 147–61). Some other special points are considered in M. Sepet, *Les Prophètes du Christ* (Paris, 1878), repr. from *Bibliothèque de l'École des Chartes* (1867, 1868, 1877); H. Anz, *Die lateinischen Magierspiele* (Leipzig, 1905); Carleton Brown, 'Caiaphas as a Palm-Sunday Prophet' (Boston, 1913, Kittredge Anniversary Papers); H. Craig, 'The Origin of the Old Testament Plays' (1913, MP x. 473–87); P. E. Kretzmann, *The Liturgical Element in the Earliest Forms of the Mediaeval Drama* (Minneapolis, 1916); W. H. Grattan Flood, 'Irish Origin of the *Officium Pastorum*' (1921, *The Month*, cxxxviii. 545–9) and 'The Irish Origin of the Easter Play' (1923, *The Month*, cxli. 349–52); R. E. Parker, 'The Reputation of Herod in Early English Literature' (1933, *Speculum*, viii. 59–67); M. H. Marshall, 'The Dramatic Tradition Established by the Liturgical Plays' (1941, *PMLA* lvi. 962–91); R. Pascal, 'On the Origins of the Liturgical Drama of the Middle Ages' (1941, *MLR* xxxvi. 369–87). Lingering traces of paganism in medieval priests are thought to have affected the liturgical plays by R. Stumpfl in *Kultspiele der Germanen als Ursprung des mittelalterlichen Dramas* (Berlin, 1936). Valuable records of liturgical plays, the texts of which are unknown, are given in V. Shull, 'Clerical Drama in Lincoln Cathedral 1318 to 1561' (1937, *PMLA* lii. 946–66). The texts of the bilingual plays from a Shrewsbury manuscript first printed by W. W. Skeat (1890, *Academy*, 11 Jan.) are edited by Young (ii. 514–23).

The *débat* on the *Harrowing of Hell* is edited by W. H. Hulme (1907, *EETS* e.s. c), the *Adam* by K. Grass (Halle, 1891) and

P. Studer (Manchester, 1918), the fragmentary Anglo-Norman *Resurrection* in L. J. N. Monmerqué and F. Michel, *Théâtre français au Moyen Âge* (Paris, 1839), and by J. G. Wright (Paris, 1931). The recently discovered Canterbury text is described in *The Times* (28 Dec. 1937). An edition by G. Cohen is believed to be-forthcoming. The Bury St. Edmunds fragment is discussed by J. P. Gilson and W. W. Greg in *The Times Literary Supplement* (1921, 26 May, 2 June). The beginnings of contemporary French drama are dealt with by G. Frank in 'The *Palatine Passion* and the Development of the Passion Play' (1920, *PMLA* xxxv. 464–83) and 'Vernacular Sources and an Old French Passion Play' (1920, *MLN* xxxv. 257–69), and by J. G. Wright in *A Study of the Themes of the Resurrection in Mediaeval French Drama* (Bryn Mawr, 1935). William de Wadington's *Manuel des Péchiez* and Robert de Brunne's *Handlyng Synne* are edited by F. J. Furnivall (1901–3, *EETS* o.s. cxix. 123), and the *Tretise of Miraclis Pleyinge* by T. Wright and J. O. Halliwell (1845, *Reliquiae Antiquae*, i. 42); E. Mätzner, *Altenglische Sprachproben* (Berlin, 1867, ii. 222); W. C. Hazlitt, *The English Drama and Stage* (1869, 73–95). *Dives et Pauper* (*c.* 1410) is discussed by H. G. Pfander (1933, *Library*, 4th Series, xiv. 299–312) and H. G. Richardson (1934, *Library*, 4th Series, xv. 31–7). The signifi-cance of the term *miraculum* is considered by G. R. Coffman in *A New Theory Concerning the Origin of the Miracle Play* (Menasha, 1914), 'The Miracle Play in England—Nomenclature' (1916, *PMLA* xxxi. 448–65), 'The Miracle Play in England' (1919, *SP* xvi. 56–66), and K. Young, 'Concerning the Origin of the Miracle Play' (Chicago, 1923, *Manly Anniversary Studies*, 254–68). An early occurrence of the term is in *A Selection of Latin Stories* (1842, Percy Soc. viii. 99).

On the social gilds L. Toulmin and J. T. Smith, *English Gilds* (1870, *EETS* o.s. xl), and H. F. Westlake, *The Parish Gilds of Mediaeval England* (1919), are valuable. W. O. Hassall, 'Plays at Clerkenwell' (1938, *MLR* xxxiii. 564–7), makes an important addition to the records of London plays in my *The Mediaeval Stage*, ii. 379–82. Those of Beverley and Lincoln are described in A. F. Leach, 'Some English Plays and Players, 1220–1458' (1901, *An English Miscellany Presented to Dr. Furnivall*, 205–34); those of Hull in A. J. Mill, 'The Hull Noah Play' (1938, *MLR* xxxiii. 489–505); and those of Scotland in A. J. Mill, *Mediaeval Plays in Scotland* (1927, St. Andrews University Publications, xxiv). The organization and methods of the Corpus Christi procession and plays are described by M. L. Spencer, *Corpus Christi Pageants in England* (New York, 1911), and their inter-relation considered by H. Craig, 'The Corpus Christi Procession and the Corpus Christi Play' (1914, *JEGP* xiii. 589–602), and L. Blair, 'A Note on the Relation of the Corpus Christi Proces-sion to the Corpus Christi Play' (1940, *MLN* lv. 2).

Selected texts of miracle plays are in T. Hawkins, *The Origin of the English Drama* (1773); J. P. Collier, *Five Miracle Plays* (1836); W. Marriott, *A Collection of English Miracle Plays or Mysteries* (Basel, 1838); W. C. Hazlitt, *Dodsley's Old Plays* (1874–6); A. W. Pollard, *English Miracle Plays, Moralities and Interludes*, with a valuable Introduction (1890, ed. 8, 1927); J. M. Manly, *Specimens of the Pre-Shakespearean Drama* (Boston, 1897) (also valuable); S. B. Hemingway, *English Nativity Plays* (New York, 1909); C. G. Child, *The Second Shepherds' Play, Everyman, and Other Early Plays* (Boston, 1910); J. S. P. Tatlock and R. G. Martin, *Representative English Plays* (New York, 1916, 1923); J. Q. Adams, *The Chief Pre-Shakespearean Dramas* (Boston, 1924).

The present writer's *The Mediaeval Stage* (1903) is more con-cerned with the organization and local distribution of miracle plays than with their dramatic and literary quality. An out-standing study is W. W. Greg, *Bibliographical and Textual Problems of the English Miracle Cycles* (1914, reprinted from *The Library*). Others are A. Ebert, 'Die englischen Mysterien' (1859, *Jahrbuch für romanische und englische Literatur*, i); R. Genée, *Die englischen Mirakelspiele und Moralitäten* (Berlin, 1878); A. R. Hohlfeld, 'Die altenglischen Kollektivmisterien' (1889, *Anglia*, xi. 219); C. Davidson, 'Studies in the English Mystery Plays' (1892, *Transactions of the Connecticut Academy of Arts and Sciences*, ix. 125–297); K. L. Bates, *The English Religious Drama* (1893); S. W. Clarke, *The Miracle Play in England* (1897); C. M. Gayley, *Plays of Our Forefathers* (New York, 1907); E. Moore, *English Miracle Plays and Moralities* (1907); W. Creizenach, 'The Early Religious Drama' (1910, *Cambridge History of English Literature*, v. 36–60); B. Cron, *Zur Entwicklungsgeschichte der englischen Misterien des Alten Testaments* (Marburg, 1913); J. M. Manly, 'The Miracle Play in Medieval England' (1927, *Transactions of the Royal Society of Literature*, vii).

Some special topics are dealt with by C. M. Gayley, 'Historical View of the Beginnings of English Comedy' (1903, *Representative*

English Comedies, vol. 1); H. Craig, 'The Origin of the Old Testament Plays' (1913, *MP* x. 473–87); E. M. Campbell, *Satire in the Early English Drama* (Columbus, Ohio, 1914); G. Frank, 'Revisions in the English Mystery Plays' (1918, *MP* xv. 565–72); G. R. Coffman, 'Corpus Christi Plays as Drama' (1929, *SP* xxvi. 411–24); F. Collins, 'Music in the Craft Cycles' (1932, *PMLA* xlvii. 613–21); B. J. Whiting, *Proverbs in the Earlier English Drama* (1938); A. J. Mill, 'Noah's Wife Again' (1941, *PMLA* lvi. 613–26).

The ultimate sources of miracle plays are discussed by C. M. Gayley (1903, op. cit.). Texts are C. von Tischendorf, *Evangelia Apocrypha* (1876, Leipzig); Peter Comestor, *Historia Scholastica* (Migne, *Patrologia Latina*, cxcviii. 1049); Jacobus de Voragine, *Legenda Aurea* (Dresden, 1846, ed. T. Graesse); R. Morris and H. Hupe, *Cursor Mundi* (1874–93, EETS o.s. lvii, lix, lxii, lxvi, lxviii, xcix, ci); R. Morris, *Legends of the Holy Rood* (1871, EETS o.s. xlvi); A. S. Napier, *The Legend of the Holy Cross* (1894, EETS o.s. ciii); F. A. Foster and W. Heuser, *The Northern Passion* (1913, 1916, 1930, EETS o.s. cxlv, cxlvii, clxxxiii, with text of *La Passion des Jongleurs*), discussed by F. A. Foster, 'The Mystery Plays and the Northern Passion' (1911, *MLN* xxvi. 169–71); W. H. Hulme, *The Gospel of Nicodemus* (1907, EETS e.s. c), discussed by W. A. Craigie, 'The Gospel of Nicodemus and the York Mystery Plays' (1901, *Furnivall Miscellany*, 52–61); *The Stanzaic Life of Christ* (1924, ed. F. A. Foster, EETS o.s. clxvi), discussed by R. H. Wilson, 'The *Stanzaic Life of Christ* and the Chester Plays' (1931, *SP* xxviii. 413–32); *The Middle English Stanzaic Versions of the Life of Saint Anne* (1927, ed. R. E. Parker, EETS e.s. clxxiv). The interaction between lyric and drama may be studied in H. Thien, *Über die englischen Marienklagen* (Kiel, 1906), G. C. Taylor, 'The English *Planctus Mariae*' (1907, *MP* iv. 605–33) and 'The Relation of the English Corpus Christi Plays to the Middle English Religious Lyric' (1907, *MP* v. 1–38), L. E. Pearson, 'Isolable Lyrics of the Mystery Plays' (1936, *ELH* iii. 228–52). G. R. Owst has called attention to the anticipation in the pulpit of the topics and manner of the plays in his valuable *Preaching in Mediaeval England* (1926) and *Literature and Pulpit in Mediaeval England* (1933). Examples are in Carleton Brown, 'An Early Mention of a St. Nicholas Play in England' (1931, *SP* xxviii. 594–601) and 'Sermons and Miracle Plays' (1934, *MLN* xlix. 394–6).

Of the Chester plays, two were edited by J. H. Markland (1818, Roxburghe Club) and the full series by T. Wright (1843–7, Shakespeare Society), and by H. Deimling and G. W. Matthews (1892, 1914, EETS e.s. lxii, cxv), a standard text. There are separate editions by W. W. Greg of the *Antichrist* (1935), of the *Fall of Lucifer* by P. E. Dustoor (Allahabad, 1930), and of recently discovered versions of *The Trial and Flagellation* by F. M. Salter and of a fragment of *The Resurrection* by W. W. Greg, in *Chester Play Studies* (1935, Malone Society). This has also texts of two versions of the Banns, and valuable dissertations by the editors. Illustrative material from civic records is in F. J. Furnivall, *The Digby Mysteries* (1882, New Shakspere Society, xviii–xxix), and R. Morris, *Chester in the Plantagenet and Tudor Reigns* (1893). Other studies are H. Ungemach, *Die Quellen der fünf ersten Chester Plays* (Erlangen, 1890); H. Utesch, *Die Quellen der Chester-Plays* (Kiel, 1909); A. C. Baugh, 'The Chester Plays and French Influence' (New York, 1923, *Schelling Anniversary Papers*); G. W. Mathews, *The Chester Miracle Plays* (Liverpool, 1925); R. H. Wilson, 'The *Stanzaic Life of Christ* and the Chester Plays' (1931, *SP* xxviii. 413–32); F. M. Salter, 'The Banns of the Chester Plays' (1939–40, *RES* xv. 432–57; xvi. 1–17, 137–48). *Vide* also *infra re* Brome *Abraham and Isaac.*

The York *Incredulity of St. Thomas* was edited from a prompter's copy by J. Croft (1797, *Excerpta Antiqua*, 103) and J. P. Collier (1859, Camden Soc. lxxiii), and the whole cycle by L. Toulmin Smith, *York Plays* (1885), a standard text. Civic records are in R. Davies, *Extracts from the Municipal Register of the City of York* (1843); M. Sellers, *The York Memorandum Book* (1912, 1915, Surtees Soc. cxx, cxxv) and *The York Mercers and Merchant Adventurers* (1918, Surtees Soc. cxxix); A. J. Mill, 'The York Bakers' Play of the Last Supper' (1935, *MLR* xxx. 145–58). Studies are O. Herrtrich, *Studien zu den York Plays* (Breslau, 1886); P. J. G. Kamann, *Über Quellen und Sprache der York Plays* (Halle, 1887); F. Holthausen, 'Beiträge zur Erklärung und Textkritik der York Plays' (1890, *Archiv*, lxxxv. 411–28) and 'Nachtrag zu den Quellen der York Plays' (1891, *Archiv*, lxxxvi. 280–2); E. Kölbing, 'Beiträge zur Erklärung und Textkritik der York Plays' (1895, *ES* xx. 179–220), K. Luick, 'Zur Textkritik der Spiele von York' (1899, *Anglia*, xxii. 384–91); F. Holthausen, 'Zur Erklärung und Textkritik der York Plays'

(1910, *ES* xli. 380–4); J. P. R. Wallis, '*Crucifixio Christi* in the York Cycle' (1917, *MLR* xii. 494–5); G. Frank, 'St. Martial of Limoges in the York Plays' (1929, *MLN* xliv. 233–5); E. Mackinnon, 'Notes on the Dramatic Structure of the York Cycle' (1931, *SP* xxviii. 433–49); M. Trusler, 'The York *Sacrificium Cayme and Abell*' (1934, *PMLA* xlix. 956–9); M. G. Frampton, 'The Brewbarret Interpolation in the York Play, the *Sacrificium Cayme and Abell*' (1937, *PMLA* lii. 895–900).

The Wakefield *Judicium* was edited by F. Douce (1822, Roxburghe Club) and the whole cycle by an uncertain editor who may have been J. Raine, J. Hunter, or J. S. Stevenson (1836, Surtees Soc.), and by G. England and A. W. Pollard (1897, EETS e.s. lxxi), a standard text. The manuscript is minutely described by L. Wann, 'A New Examination of the Manuscript of the Towneley Plays' (1928, *PMLA* xliii. 137–52). Its evidence for an origin at Wakefield is noted and confirmed by proof of a cycle there by W. W. Skeat, 'The Locality of the Towneley Plays' (1893, 2 Dec., *Athenaeum*); M. H. Peacock (1900, *Yorkshire Archaeological Journal*, xv; 1901, *Anglia*, xxiv. 509–24; 1925, *TLS* 5 March; 1928, *TLS* 7 June), and J. W. Walker (*Yorks Archaeological Soc. Record Series*, lxxiv. 20–2). The growth of Wakefield is described in J. W. Walker, *Wakefield, its History and People* (1934). Studies of the cycle are A. Bunzen, *Ein Beitrag zur Kritik der Wakefielder Mysterien* (Kiel, 1903); F. W. Cady, 'The Liturgical Basis of the Towneley Mysteries' (1909, *PMLA* xxiv. 419–69), 'The Couplets and Quatrains in the Towneley Mystery Plays' (1911, *JEGP* x. 572–84), 'The Wakefield Group in Towneley' (1912, *JEGP* xi. 244–62), 'The Passion Group in Towneley' (1913, *MP* x. 587–600); E. F. Williams, *The Comic Element in the Wakefield Mysteries* (Berkeley, 1914); Carleton Brown, 'The Towneley *Play of the Doctors* and the *Speculum Christiani*' (1916, *MLN* xxxi. 223–6); A. S. Cook, 'Another Parallel to the Mak Story' (1916, *MP* xiv. 11–15); A. C. Baugh, 'The Mak Story' (1918, *MP* xv. 729–34); F. Holthausen, 'Studien zu den Towneley Plays' (1924, *ES* lviii. 161–78); M. Carey, *The Wakefield Group in the Towneley Cycle* (Göttingen, 1929). On the Wakefield Master in particular are M. G. Frampton, 'The Date of the Flourishing of the "Wakefield Master"' (1935, *PMLA* l. 631–60 and 1938, *ibid.* liii. 86–117) and 'The Processus Talentorum' (1944, *ibid.* lix. 646–54); M. Trusler, 'The Language of the Wakefield Play-

wright' (1936, *SP* xxxiii. 15–39); J. H. Smith, 'Another Allusion
to Costume in the Work of the "Wakefield Master" ' (1937,
PMLA lii. 901–2). A suggested identification of the Master with
one Gilbert Pilkington, only known as a scribe, by O. Cargill,
'Authorship of the *Secunda Pastorum*' (1926, *PMLA* xli. 810–31),
is confuted by F. A. Foster, 'Was G.P. Author of the *Secunda
Pastorum?*' (1928, *PMLA* xliii. 124–36) and M. G. Frampton,
'G.P. Once More' (1932, *PMLA* xlvii. 622–35).

The Wakefield borrowings from York and the theory of a
'Parent-Cycle' suggested by Davidson (op. cit.) are considered
in H. E. Coblentz, 'A Rime-Index to the Parent Cycle' (1895,
PMLA x. 487–557); M. C. Lyle, *The Original Identity of the York
and Towneley Cycles* (Minneapolis, 1919); E. G. Clark, 'The
York Plays and the *Gospel of Nichodemus*' (1928, *PMLA* xliii,
153–61); G. Frank, 'On the Relation between the York and
Towneley Plays' (1929, *PMLA* xliv. 313–19); M. C. Lyle, 'The
Original Identity of the York and Towneley Cycles—a Rejoin-
der' (1929, *PMLA* xliv. 319–28); F. W. Cady, 'Towneley, York,
and True-Coventry' (1929, *SP* xxvi. 386–400); C. G. Curtiss,
'The York and Towneley Plays on *The Harrowing of Hell*' (1933,
SP xxx. 24–33); J. H. Smith, 'The Date of Some Wakefield
Borrowings from York' (1938, *PMLA* liii. 595–600), M. G.
Frampton, 'The Towneley *Harrowing of Hell*' (1941, *PMLA*
lvi. 105–19), and 'Towneley xx: the *Conspiracio (et Capcio)*'
(1943, *PMLA* lviii. 920–37).

The Coventry Shearmen and Tailors' play was edited by T.
Sharp in *Illustrative Papers of the History of Coventry* (1817), and
with matter from records in *A Dissertation on the Coventry Mysteries*
(1825), the Weavers Play was edited by J. B. Gracie (1836,
Abbotsford Club), and both plays, with many further records,
in H. Craig's standard edition (1902, EETS e.s. lxxxvii). Useful
supplements are M. D. Harris, *Life in an English Town* (1898)
and *The Coventry Leet Book* (1907–13, EETS o.s. cxxxiv, cxxxv,
cxxxviii, cxlvi).

Five plays from the *Ludus Coventriae* were edited in W. Dugdale, *Monasticon Anglicanum* (1817–30, vi, pt. 3, 1534), the
whole by J. O. Halliwell (1841, Shakespeare Society) and K. S.
Block (1922, EETS e.s. cxx), a standard edition, and *The
Assumption* by W. W. Greg (1915). Studies are M. Kramer,
Sprache und Heimath des sogenannten 'Ludus Coventriae' (Halle,
1892); E. N. S. Thompson, 'The *Ludus Coventriae*' (1906, *MLN*

xxi. 18–20); E. L. Swenson, *An Inquiry into the Composition and Structure of Ludus Coventriae* (Minneapolis, 1914); M. H. Dodds, 'The Problem of the *Ludus Coventriae*' (1914, *MLR* ix. 79–91); H. R. Patch, 'The *Ludus Coventriae* and the Digby *Massacre*' (1920, *PMLA* xxxv. 324–43); H. Hartman, 'The Home of the *Ludus Coventriae*' (1926, *MLN* xli. 530–1); G. C. Taylor, 'The *Christus Redivivus* of Nicholas Grimald and the Hegge Resurrection Plays' (1926, *PMLA* xli. 840–59).

The *Innocents* from the Digby MS. was edited by T. Hawkins, *Origin of the English Drama* (1773), and again with the *Conversion of St. Paul* and the *Mary Magdalen*, by T. Sharp (1835, Abbotsford Club), and the whole MS. by F. J. Furnivall, *The Digby Mysteries* (1882, New Shakspere Society, repr. 1896, EETS e.s. lxx), a standard text. Studies are K. Schmidt, *Die Digby-Spiele* (Berlin, 1884); H. R. Patch, 'The *Ludus Coventriae* and the Digby *Massacre*' (1920, *PMLA* xxxv. 324–43); J. R. Moore, 'Miracle Plays, Minstrels, and Jigs' (1933, *PMLA* xlviii. 943–5). *The Burial and Resurrection* was edited by T. Wright and J. O. Halliwell (1843, *Reliquiæ Antiquae*, ii). Plays on saints are discussed by J. M. Manly, 'The Miracle Play in Mediaeval England' (1927, *Transactions of the Royal Society of Literature*, vii).

The Newcastle *Noah's Ark* was edited by H. Bourne (1736, *History of Newcastle*), J. Brand (1789, *History of Newcastle*), T. Sharp, *Dissertation on the Coventry Mysteries* (1825), F. Holthausen (1897, *Göteborg's Högskola's Årsskrift*), R. Brotanek (1899, *Anglia*, xxi. 165–200); the Norwich *Adam and Eve* by R. Fitch (1856, *Norfolk Archaeology*, v); the Brome *Abraham and Isaac* by L. T. Smith (1884, *Anglia*, vii. 316–37) and W. Rye (1887, *Norfolk Antiquarian Miscellany*, iii); the Dublin *Abraham and Isaac* by J. P. Collier, *Five Miracle Plays* (1836), and R. Brotanek (1899, *Anglia*, xxi. 21–55); the Croxton *Sacrament* by W. Stokes (1861, *Transactions of the Philological Society*). All are brought together in O. Waterhouse, *The Non-Cycle Mystery Plays* (1909, EETS e.s. civ), a standard text.

The Brome play is studied, in relation to the Chester *Abraham and Isaac*, by A. R. Hohlfeld, 'Two Old English Mystery Plays on the Subject of Abraham's Sacrifice' (1890, *MLN* v. 111–19), C. A. Harper, 'Comparison of the Brome and Chester Plays of Abraham and Isaac' (Boston, 1910, *Studies Presented to Agnes Irwin*), M. D. Fort, 'Metres of the Brome and Chester Abraham and Isaac Plays' (1926, *PMLA* xli. 832–9), P. E. Dustoor,

Textual Notes on the Chester MSS. and the Brome Play (Allahabad, 1930).

The morality plays are discussed in J. Bolte, 'Der Teufel in der Kirche' (1897, *Zeitschrift für vergleichende Litteraturgeschichte*, xi); L. W. Cushman, *The Devil and the Vice in English Dramatic Literature before Shakespeare* (Halle, 1900); E. Eckhardt, *Die lustige Person im älteren englischen Drama* (Berlin, 1902); E. N. S. Thompson, 'The English Moral Plays', a valuable study (1910, *Transactions of the Connecticut Academy of Arts and Sciences*, xiv. 291–414); E. J. Haslinghuis, *De Duivel in het Drama der Middeleeuwen* (Leyden, 1912); W. R. Mackenzie, *The Origin of the English Morality* (Washington, 1915); J. R. Moore, 'Ancestors of Autolycus in the English Moralities and Interludes' (*Washington Univ. Studies* IX, 1922); T. E. Allison, 'The Paternoster Play and the Origin of the Vice' (1924, *PMLA* xxxix. 789–804); E. Eckhardt, 'Die metrische Unterscheidung von Ernst und Komik in den englischen Moralitäten' (1927, *ES* lxii. 152–69); W. Farnham, *The Mediaeval Heritage of Elizabethan Tragedy* (Boston, 1936), a useful survey. The debate in Heaven is studied by H. Traver, 'The Four Daughters of God' (1925, *PMLA* xl. 44–92), and the theme of Death by E. P. Hammond, *English Verse between Chaucer and Surrey* (Durham, N. Carolina, 1927) and F. Warren, *The Dance of Death* (1929, EETS o.s. clxxxi). K. Young collects 'Records of the York Paternoster Play' (1932, *Speculum*, vii. 540–6).

There is no complete collection of morality texts. F. J. Furnivall and A. W. Pollard edited *The Macro Plays* (1904, EETS e.s. xci), a standard text which has the *Castle of Perseverance*, *Wisdom*, and *Mankind*. These three, with *Everyman*, are also in J. S. Farmer, *Tudor Facsimile Texts* (1907, 1908, 1912). *The Pride of Life* is edited by J. Mills (1891, *Proc. Royal Society of Antiquaries of Ireland*); A. Brandl, *Quellen des weltlichen Dramas in England vor Shakespeare* (Strassburg, 1898, *Quellen und Forschungen*, lxxx), a standard text, and F. Holthausen (1901, *Archiv*, cviii. 32–59). A study is Carleton Brown, 'The *Pride of Life* and the Twelve Abuses' (1912, *Archiv*, cxxviii. 72–8). W. K. Smart discusses *The Castle of Perseverance* (Chicago, 1923, *Manly Anniversary Studies*, 42–53). *Wisdom* was edited by T. Sharp in *Ancient Mysteries* (1835, Abbotsford Club) from an incomplete text in the Digby MS. which is also collated by Furnivall and Pollard (*supra*), and from the Macro MS., by W. B. D. Turnbull

(1837, Abbotsford Club). A study is W. K. Smart, *Some English and Latin Sources and Parallels for the Morality of Wisdom* (Madison, 1912). *Mankind* is edited by A. Brandl (op. cit.). Studies are M. M. Keiller, 'The Influence of *Piers Plowman* on *Mankind*' (1911, *PMLA* xxvi. 339–55); W. R. Mackenzie, 'A New Source for *Mankind*' (1912, *PMLA* xxvii. 98–105); W. K. Smart, 'Some Notes on *Mankind*' (1916, *MP* xiv. 45–58) and '*Mankind* and the Mumming Plays' (1917, *MLN* xxxii. 21–5). The early prints of *Everyman* by Pynson and Skot are well edited by W. W. Greg (Louvain, 1904, 1909, 1910, *Materialien zur Kunde des älteren englischen Dramas*, iv, xxiv, xxviii). Other editions are by T. Hawkins (1773, *Origin of the English Drama*, i), K. Goedeke (Hanover, 1865), W. C. Hazlitt (1874, *Dodsley's Old English Plays*, i), H. Logeman (Ghent, 1892, with *Elckerlijk*), K. H. de Raaf (Groningen, 1897, with *Elckerlijk*), M. J. Moses (New York, 1903). Studies are by H. Logeman (Ghent, 1902); A. Roersch, 'Elckerlijk–Everyman' (1904, *Archiv*, cxiii. 13–16); W. Bang, 'Zu Everyman' (1905, *ES* xxxv. 444–9); J. M. Manly, '*Elckerlijk–Everyman*: The Question of Priority' (1910, *MP* viii. 269–77). The *Processus Satanae* is edited by W. W. Greg (1931, *Malone Society Collections*, ii. 239–50).

The probability of secular plays by *mimi* is considered in A. Nicoll, *Masks, Mimes and Miracles* (1931). The *Interludium de Clerico et Puella* is edited by T. Wright and J. O. Halliwell (1841, *Reliquiae Antiquae*, i), W. Heuser, 'Das I. d. C. et P. und das Fabliau von Dame Siriz' (1907, *Anglia*, xxx. 306–19), and E. K. Chambers (1903, *Mediaeval Stage*, ii. 324); and *Dux Moraud* by W. Heuser (1907, *Anglia*, xxx. 180–208). For the plays on Robin Hood, *vide* Bibliography to Chapter III.

CHAPTER II. THE CAROL AND THE LYRIC

The *Index of Middle English Verse* cited above largely supersedes Carleton Brown, *A Register of Middle English Religious and Didactic Verse* (1916–20, Bibliographical Soc.).

Lydgate's *Minor Poems* are well edited by H. N. MacCracken (1910, 1933, EETS e.s. cvii; o.s. cxcii). Other texts of religious verse are in F. J. Furnivall, *Political, Religious and Love Poems* (1866, EETS o.s. xv, re-edited 1903), and *Hymns to the Virgin and Christ* (1866, EETS o.s. xxiv); G. G. Perry, *Religious Pieces in Prose and Verse from R. Thornton's MS.* (1867, 1913, EETS o.s.

xxvi); J. R. Lumby, *Ratis Raving and other Moral and Religious Pieces* (1870, EETS o.s. xliii); M. Day, *Poems of the Wheatley MS.* (1917, EETS o.s. clv). But a large number of non-carol poems remain scattered. C. Horstmann and F. J. Furnivall, *Minor Poems of the Vernon MS.* (1892, 1901, EETS o.s. xcviii, cxvii), is valuable for comparison with later refrain poems. Examples of religious poems, carol and non-carol, are in anthologies. The most comprehensive is Carleton Brown, *Religious Lyrics of the Fifteenth Century* (1939). Others are F. A. Patterson, *The Middle English Penitential Lyric* (New York, 1911), and Lord David Cecil, *The Oxford Book of Christian Verse* (1940). Both religious and secular pieces are in J. Ritson, *A Select Collection of English Songs* (1783) and *Ancient Songs and Ballads* (pr. 1787, dated 1790, publ. 1792, ed. W. C. Hazlitt, 1877); F. T. Palgrave, *The Golden Treasury* (1861 and later editions) and *A Treasury of Sacred Song* (1889); A. T. Quiller-Couch, *The Oxford Book of English Verse* (1900, new ed. 1939); E. K. Chambers and F. Sidgwick, *Early English Lyrics* (1907).

From the voluminous literature, mainly French, on the *caroles* and related poetry may be noted as of special value G. Paris, *Études sur la Poésie lyrique en France aux douzième et troisième Siècles* (Paris, 1887–8, École des Hautes-Études) and *La Littérature Française au Moyen Âge* (Paris, 1888, ed. 5, 1914); L. Petit de Julleville, *Histoire de la langue et de la littérature françaises*, with chapter on *Les Chansons* by A. Jeanroy (Paris, 1896–1900); A. Jeanroy, *Les Origines de la poésie lyrique en France au Moyen Âge* (Paris, 1889, 1904, 1925). Useful collections of texts are in K. Bartsch, *Romanzen und Pastourellen* (Leipzig, 1870); A. Jeanroy, *Chansons, Jeux Parties et Refrains Inédits du xiiie siècle* (Paris, 1902); J. Brakelmann, *Les Plus Anciens Chansonniers Français* (Paris and Marburg, 1870–96); G. Raynaud et H. Lavoix, *Recueil de Motets Français des xiie et xiiie Siècles* (Paris, 1882–4); K. Bartsch, *Chrestomathie Provençale* (Elberfeld, 1903). For the *Carmina Burana* of the *Scholares Vagantes*, *vide* Bibliography for Ch. I. On *Bele Aelis* may be consulted A. Lecoy de la Marche, *La Chaire française au Moyen Âge* (Paris, 1886); G. Paris, '*Bele Aelis*' (in *Mélanges dédiés à Carl Wahlund* (Paris, 1896); G. Schläger, 'Sur la musique et la construction strophique des romances françaises' (Halle, 1900, *Forschungen zur romanischen Philologie, Festgabe für H. Suchier*); R. Meyer, J. Bédier, et P. Aubry, *La Chanson de Bele Aelis, par le Trouvère Baude de la Quarière* (Paris, 1904). Good modern editions of romances containing *refrains* are G. Servois, *Gui de Dôle* (Paris, 1893, with a study of *Les Chansons* by G. Paris); P. Lejeune-Dehousse, *Gui de Dôle* (Paris, 1936); F. Michel, *La Viollette* (Paris, 1834); D. L. Buffum, *La Viollette* (Paris, 1928); A. Curvers, *Renart* (Paris, 1937). Special studies are J. Bédier, 'Les Plus Anciennes Danses Françaises' (1906, *Revue des Deux Mondes*, xxxi. 398); L. Jordan, 'Der Reigentanz Carole und seine Lieder' (1931, *Zeitschrift für romanische Philologie*, li); P. Verrier, 'La plus vieille Citation de Carole' (1932, *Romania*, lviii).

On the relation of early dance and song to the *observatio paganorum* and on the secular element in Christmas, I may refer to bk. ii of my *The Mediaeval Stage* (1903), and on the former also to my essay on 'Some Aspects of Mediaeval Lyric' in *Early English Lyrics* (1907) and to F. Liebermann, 'Zu Liedrefrain und Tanz im englischen Mittelalter' (1920, *Archiv*, cxl. 261-2). The Dancers of Kölbigk are well studied by E. Schröder, 'Die Tänzer von Kölbigk' (1897, *Zeitschrift für Kirchengeschichte*, xvii. 94-164); G. Paris, 'Les Danseurs maudits' (1899, *Journal des Savants*, 733-46); W. P. Ker, 'On the Danish Ballads' (1904-8, *Scottish Historical Review*, i. 357-78; v. 385-401); K. Sisam, *Fourteenth Century Verse and Prose* (1921, 1933); and from the point of view of balladry by G. H. Gerould, *The Ballad of Tradition* (1932).

English refrain poems which lead up to the carols may be profitably studied in K. Böddeker, *Altenglische Dichtungen des MS. Harleian 2253* (Berlin, 1878); W. Heuser, *Kildare Gedichte* (1904, *Bonner Beiträge zur Anglistik*, xiv); E. K. Chambers and F. Sidgwick, *Early English Lyrics* (1907); K. Sisam, *Fourteenth Century Verse and Prose* (1921, 1933); C. Brown, *English Lyrics of the Thirteenth Century* (1932) and *Religious Lyrics of the Fourteenth Century* (1924); R. H. Robbins, 'The Earliest Carols and the Franciscans' (1938, *MLN* liii. 239-45) and 'The Authors of the Middle English Religious Lyrics' (1940, *JEGP* April).

The fullest collection of carols is the admirable one by R. L. Greene, *The Early English Carols* (1935), which has a long introduction and much bibliographical matter. Others, sometimes modernized or adapted for devotional singing or including 'traditional' texts, are G. Davies, *Some Ancient Christmas Carols* (1822, 1823); W. Sandys, *Christmas Carols, Ancient and Modern* (1833); T. Wright, *Specimens of Old Christmas Carols* (1841, Percy

Soc.); E. F. Rimbault, *A Little Book of Christmas Carols* (1846) and *A Collection of Old Christmas Carols* (1863); E. Sedding, *A Collection of Antient Christmas Carols* (1860, 1863); J. Sylvester, *A Garland of Christmas Carols* (1861); W. H. Husk, *Songs of the Nativity* (1868); A. H. Bullen, *A Christmas Garland* (1885); C. J. Sharp, *Folk-Song Carols* (1913); E. Rickert, *Ancient English Christmas Carols, 1400–1700* (1914); P. Dearmer, R. Vaughan Williams, and Martin Shaw, *The Oxford Book of Carols* (1928). The principal manuscripts which contain carols have received separate treatment, which often extends to non-carol poems found with them. Editions and studies are: (*a*) (Sloane MS. 2593), T. Wright, *Songs and Carols* (1836); T. Wright, *Songs and Carols* (1856, Warton Club); B. Fehr, 'Die Lieder der HS. Sloane 2593' (1902, *Archiv*, cix. 33–72). (*b*) (Bodleian Douce MS. 302), J. O. Halliwell, *The Poems of John Audelay* (1844, Percy Soc.); E. K. Chambers and F. Sidgwick, 'Fifteenth Century Carols by John Audelay' (1910–11, *MLR* v. 473–91, vi. 68–84); E. K. Whiting, *The Poems of John Audelay* (1931, EETS o.s. clxxxiv), a standard text; R. Priebsch, 'John Audelay's Poem on the Observance of Sunday and its Source' (1901, *Furnivall Miscellany*, 397–407); W. F. Storck and R. Jordan, 'John Awdelays Gedicht *De Tribus Regibus Mortuis*' (1910–11, *ES* xliii. 177–88); J. K. Ramussen, *Die Sprache John Audelays* (Bonn, 1914). (*c*) (Trin. Coll. Camb. MS. O.3.58), J. A. Fuller-Maitland and W. S. Rockstro, *English Carols of the Fifteenth Century* (1891). (*d*) (Bodl. Selden MS. B.26), Sir John Stainer, *Early Bodleian Music* (1901); F. M. Padelford, 'English Songs in MS. Selden B.26' (1912, *Anglia*, xxxvi. 79–115); Sir R. Terry, *A Medieval Carol Book* (1931). (*e*) (Bodl. MS. Eng. Poet. e.1), T. Wright, *Songs and Carols* (1847, Percy Soc.). (*f*) (Camb. St. John's MS. S.54), M. R. James and G. C. Macaulay, 'Fifteenth Century Carols and Other Pieces' (1913, *MLR* viii. 68–87). (*g*) (Camb. Univ. MS. Ee.1.12), J. Zupitza, 'Die Gedichte des Franziskaners Jakob Ryman' (1892, *Archiv*, lxxxix. 167–338). (*h*) (Balliol MS. 354), E. Flügel, 'Englische Weihnachtslieder aus einer Handschrift des Balliol College zu Oxford' (Leipzig, 1894, *Festgabe für Rudolf Hildebrand*), and 'Liedersammlungen des xvi Jahrhunderts besonders aus der Zeit Heinrichs viii' (1903, *Anglia*, xxvi. 94–285); A. W. Pollard, 'English Carols' (1903, *Fifteenth Century Prose and Verse*, 81–96); R. Dyboski, *Songs, Carols and Other Miscellaneous Poems from Balliol MS. 354*

(1908, EETS e.s. ci).　　(*i*) (B.M. Addl. MS. 5465), B. Fehr, 'Die Lieder des Fairfax MS.' (1901, *Archiv*, cvi. 48).　　(*k*) (B.M. Addl. MS. 5665), B. Fehr, 'Die Lieder der HS. Add. 5665' (1901, *Archiv*, cvi. 262).　　(*l*) (B.M. Addl. MS. 31922), E. Flügel, 'Liedersammlungen' *ut supra* (1889, *Anglia*, xii. 226–56).　　(*m*) (B.M. Royal MS. App. 58), E. Flügel, 'Liedersammlungen' *ut supra* (1889, *Anglia*, xii. 256–72).　　E. B. Reed, *Christmas Carols Printed in the Sixteenth Century* (Cambridge, Mass., 1932) reproduces the texts in facsimile, with a valuable commentary. His *English Lyrical Poetry from its Origins to the Present Time* (New Haven, 1912) takes a wider sweep.

On the Tudor musical development, which affected the carol, may be studied J. Stafford Smith, *Musica Antiqua* (1812); W. Nagel, 'Annalen der englischen Hofmusik' (1894–5, *Beitrag zu den Monatshefte für Musikgeschichte*, xxvi); W. H. Hadow, *The Oxford History of Music*, vols. i, ii (1901–5, 1929–32); W. H. Grattan Flood, *Early Tudor Composers* (1925). On the relation of English to Latin song, P. S. Allen, 'Mediaeval Latin Lyrics' (1908–9, *MP* v. 423–76; vi. 3–43, 137–40, 385–406); F. J. E. Raby, *History of Christian Latin Poetry* (1927); W. O. Wehrle, *The Macaronic Hymn Tradition in Medieval English Literature* (Washington, 1933). And on some special features H. Thien, *Über die mittelenglischen Marienklagen* (Kiel, 1906); G. C. Taylor, 'The English *Planctus Mariae*' (1907, *MP* iv. 605–37); H. E. Sandison, *Chanson d'Aventure in Middle English* (1913, Bryn Mawr Monographs, xii); J. M. Berdan, '*The Scholastic Tradition*' in Early *Tudor Poetry* (1931); E. Tilgner, *Die Aureate Terms als Stilelement bei Lydgate* (Berlin, 1936).

Political poems are collected by F. Madden, 'Political Poems of the Reigns of Henry VI and Edward IV' (1841–2, *Archaeologia*, xxix); T. Wright, *Political Poems and Songs* (1859–61, Rolls Series); J. Kail, *Political and Other Poems* (1904, EETS o.s. cxxiv, incomplete). *The Brut, or The Chronicle of England*, edited by F. W. D. Brie (1906–8, EETS o.s. cxxxi, cxxxvi) is not itself lyrical, although based in part on popular song. A valuable survey is C. L. Kingsford, *English Historical Literature in the Fifteenth Century* (1913). It may be supplemented by F. J. Starke, *Populäre englische Chroniken des 15 Jahrhunderts* (Berlin, 1905), and on literary and social conditions by J. Gairdner, *The Paston Letters* (1872–5, 1900–1); H. S. Bennett, *The Pastons and their England* (1932) and 'The Author and his Public in the Fourteenth and

Fifteenth Centuries' (1938, *Essays and Studies by Members of the English Association*, xxiii).

Secular poems are scattered among carol collections and modern anthologies. *London Lickpenny*, now generally rejected from the Lydgate canon, is in E. Arber, *The Dunbar Anthology* (1901), 113, and E. P. Hammond, *English Verse between Chaucer and Surrey* (Durham, N. Carolina, 1927), 237. *Syr Penny* is in W. C. Hazlitt's old-fashioned but still useful *Remains of the Early Popular Poetry of England* (1864–6), i. 161, iii. 82. The *Reliquiae Antiquae* were collected by T. Wright and J. O. Halliwell (1841–3). *The Lover's Mass* is edited by E. P. Hammond (1908, *JEGP* vii) and in *English Verse, ut supra*. *The Nut-Brown Maid* was printed by T. Wright from Arnold's text in 1836, and that version has generally been followed in later collections. The Balliol text does not seem to have been separately edited, but it is conflated with Arnold's in E. K. Chambers and F. Sidgwick, *Early English Lyrics* (1907), 34.

CHAPTER III. POPULAR NARRATIVE POETRY AND THE BALLAD

The relation between narrative and lyric in poetry is well discussed in W. P. Ker, *Form and Style in Poetry* (1928).

The Agincourt poems are in N. H. Nicolas, *History of the Battle of Agincourt* (1827, 1832, 1833, pp. 301 and App. 69), a.nd W. C. Hazlitt, *Remains of the Early Popular Poetry of England* (1866, ii. 93); that by John Page on Rouen in J. Gairdner, *Collections of a London Citizen* (1876, Camden Soc.), and W. C. Hazlitt (*ut supra*, ii. 92); that on Calais in F. Madden, 'Political Poems' (1842, *Archaeologia*, xxxiii. 129–32). The *Brut* is edited by F. W. D. Brie (1906–8, EETS o.s. cxxxi, cxxxvi), the *Chronicles of London* by C. L. Kingsford (1905), and *The Great Chronicle of London* by E. H. Dring (1913), and in a fuller text, from a recently discovered MS., by A. H. Thomas and I. D. Thornley, *The Great Chronicle of London* (1938). A valuable study is again C. L. Kingsford, *English Historical Literature in the Fifteenth Century* (1913).

Tales of the King and the Subject and other edifying or amusing narratives, not always in very reliable texts, are dispersed over J. Ritson, *Ancient Popular Poetry* (1791), and *Ancient Songs and Ballads* (pr. 1787, dated 1790, publ. 1792, ed.

W. C. Hazlitt, 1877); H. Weber, *Metrical Romances* (1810); E. V. Utterson, *Select Pieces of Early Popular Poetry* (1817); D. Laing, *Select Remains of the Early Popular Poetry of Scotland and the Border* (1822, ed. W. C. Hazlitt, 1895); C. H. Hartshorne, *Ancient Metrical Tales* (1829); T. Wright, *Early English Poetry* (1836); J. O. Halliwell and T. Wright, *Reliquiae Antiquae* (1841, 1845); J. O. Halliwell, *Contributions to Early English Literature* (1849) and *Early English Miscellanies* (1855, Warton Club, from Porkington MS. 10); W. C. Hazlitt, *Remains of the Early Popular Poetry of England* (1864-6). *John the Reeve* is in the Percy Folio MS. edited by J. W. Hales and F. J. Furnivall (1867-8, ii. 550). There are separate editions of *How A Merchant did his Wife Betray* by D. Laing (1857, Abbotsford Club), *The Merchant and his Son* by C. Hopper (1859, Camden Soc.), Adam of Cobsam's *The Wright's Chaste Wife* by F. J. Furnivall and W. A. Clouston (1865, 1886, EETS o.s. xii, lxxxiv), *The Childe of Bristowe* by C. Horstmann (Heilbronn, 1881, *Altenglische Legenden*, 314), *Rauf Coilȝear* by S. J. Herrtage (1882, EETS e.s. xxxix), M. Toundorf (Berlin, 1894), and F. J. Amours in *Scottish Alliterative Poems* (1897, *STS* lxxxii), *King Edward and the Hermit* by A. Kurz (Erlangen, 1905). Some late versions of King and Subject narratives are in Percy's *Reliques* (*v. infra*).

A full bibliography of broadsides would be out of place here. Important are A. Clark, *The Shirburn Ballads* (1907); C. H. Firth, 'Ballads and Broadsides', in *Shakespeare's England* (1917, 1926, with additions) and *Essays Historical and Literary* (1938); H. E. Rollins, 'The Black-Letter Broadside Ballad' (1919, *PMLA* xxxiv. 258-339), 'Martin Parker' (1919, *MP* xvi. 449-74) and 'Concerning Bodleian MS. Ashmole 48' (1919, *MLN* xxxiv. 340-51, on Richard Sheale), 'Analytical Index to the Ballad-Entries (1557-1709) in the Registers of the Company of Stationers of London' (1924, *SP* xxi. 1-324); E. von Schaubert, 'Zur Geschichte der Black-Letter Broadside Ballad' (1926, *Anglia*, l. 1-61). Thomas Deloney's writings are edited by F. O. Mann (1912). *Ane Compendious Buik of Godlie Psalms and Spiritual Sangs* (1567), often called *The Gude and Godlie Ballats*, was probably compiled by the brothers James, John, and Robert Wedderburn. A modern edition is by A. F. Mitchell (1897, STS). Of *The Complaynt of Scotland* (1548) are editions by John Leyden (1801) and J. A. H. Murray (1872, EETS e.s. xvii, xviii). The passage on tales, songs, and dances, with Captain Cox's ballads in 1575, is given in F. J. Furnivall, *Robert Laneham's Letter* (1871, Ballad Soc.; 1907, Shakespeare Library). Ambrose Philips (*c.* 1675–1749) is the conjectured compiler of *A Collection of Old Ballads* (1723–5, type-facsimile, 1866). The Percy Folio MS. (*c.* 1650) was edited by J. W. Hales and F. J. Furnivall (1867–8). Thomas Percy's *Reliques of Ancient English Poetry* was published in 1765. Later editions are of 1767, 1775, and 1794. A more recent one is by M. M. A. Schröer (Heilbronn, 1889–93). Percy's work and the theories of *Das Volk dichtet*, which it influenced, are discussed in H. A. Beers, *History of Romanticism in the Eighteenth Century* (1899) and *History of Romanticism in the Nineteenth Century* (1902), S. B. Hustvedt, *Ballad Criticism in Scandinavia and Great Britain during the Eighteenth Century* (New York, 1916), and L. Dennis, 'Thomas Percy: Antiquarian *versus* Man of Taste' (1942, *PMLA* lvii. 140–54).

Collections by publishers, containing a few Scottish ballads, probably from broadsides, are Allan Ramsay, *The Tea-Table Miscellany* (1724–7, other eds. to 1763); *The Hive* (1724, 1732–3); W. Thomson, *Orpheus Caledonicus* (1725, 1733); *One Hundred and Fifty Scots Songs* (1768, London); G. Caw, *The Poetical Museum* (1784, Hawick); J. Johnson, *The Scots Musical Museum* (1787–1803), largely edited by Burns. A few separate texts were issued in 1755 by R. and A. Foulis of Glasgow. More important are the gatherings from Scottish oral tradition in David Herd, *The Ancient and Modern Scots Songs, Heroic Ballads, etc.* (1769, 1776, 1791, 1869); John Pinkerton, *Scottish Tragic Ballads* (1781) and *Select Scotish Ballads* (1783); Joseph Ritson, *The Caledonian Muse* (1785, burnt before publication, 1821) and *Scotish Song* (1794); Sir Walter Scott, *Minstrelsy of the Scottish Border* (1802–3, 1806, 1810, ed. J. G. Lockhart, 1833, ed. T. F. Henderson, 1902); Robert Jamieson, *Popular Ballads and Songs* (1806); John Finlay, *Scottish Historical and Romantic Ballads* (1808); Robert Cromek, *Select Scottish Songs* (1808), with notes by Burns, and *Remains of Nithsdale and Galloway Song* (1810); Alexander Laing, *Scarce Ancient Ballads* (1822) and *The Thistle of Scotland* (1823); Charles Kirkpatrick Sharpe, *A Ballad Book* (1823, repr. 1880, 1883); James Maidment, *A North Countrie Garland* (1824, 1884), *Reliquiae Scoticae* (1828), *A New Book of Old Ballads* (1844) and *Scottish Ballads and Songs* (1859); Allan Cunningham, *The Songs of Scotland, Ancient and Modern* (1825); Peter Buchan, *Gleanings of Scotch, English, and*

Irish Ballads (Peterhead, 1825), *Ancient Ballads and Songs of the North of Scotland* (1828), and *Scottish Traditional Versions of Ancient Ballads* (1845, ed. J. H. Dixon for Percy Soc.); Robert Chambers, *Popular Rhymes of Scotland* (1826, 1870), *Twelve Romantic Scottish Ballads* (1844) and *The Romantic Scottish Ballads* (1859); William Motherwell, *Scottish Minstrelsy, Ancient and Modern* (1827); George Kinloch, *Ancient Scottish Ballads* (1827).

A standard collection is F. J. Child, *The English and Scottish Popular Ballads* (Boston, 1857–9, 1882–98), which contains all the versions known to the compiler of 305 ballads. On this was based H. C. Sargent and G. L. Kittredge, *English and Scottish Popular Ballads* (Boston, 1904, London, 1905), with one or more versions of each of Child's ballads except five 'high-kilted' ones, a valuable introduction by Professor Kittredge, and a bibliographical appendix, giving *inter alia* the repositories of many manuscript ballads in the form in which they were originally taken down from singing or recitation. Other collections and anthologies are: Thomas Evans, *Old Ballads, Historical and Narrative* (1777–1784, ed. R. H. Evans, 1810); A. Whitelaw, *The Book of Scottish Ballads* (1845); J. H. Dixon, *Ancient Poems, Ballads and Songs of the Peasantry of England* (1846, Percy Soc.); R. Bell, *Early Ballads* (1856) and *Ancient Poems, Ballads and Songs of the Peasantry of England* (1857); W. E. Aytoun, *The Ballads of Scotland* (1858, 1859, 1870); W. Allingham, *The Ballad Book* (1864); J. Collingwood Bruce and J. Stokoe, *Northumbrian Minstrelsy* (1882); A. Lang, 'Ballads' (1880, in T. H. Ward, *The English Poets*, i. 203–47), *Border Ballads* (1895), and *A Collection of Ballads* (1897); G. R. Tomson, *Ballads of the North Countrie* (1888); F. B. Gummere, *Old English Ballads* (Boston, 1894); G. Eyre-Todd, *Ancient Scots Ballads* (1895); F. Sidgwick, *Popular Ballads of the Olden Time* (1903–1912); A. Quiller-Couch, *The Oxford Book of Ballads* (1910); W. M. Hart, *English Popular Ballads* (Chicago, 1916); G. H. Stempel, *A Book of Ballads* (1917); J. Goss, *Ballads of Britain* (1937).

Many versions of Aberdeenshire ballads, additional to Child's, are in Gavin Greig's 'Folk-Song of the North-East' (1909, 1914, *Buchan Observer*), and his *Last Leaves of Traditional Ballads and Airs* (1925, ed. A. Keith). An important new text of *King John and the Bishop* (Child No. 45) from an English MS. of *c.* 1550–70 is contributed by R. D. Cornelius (1931, *PMLA* xlvi. 1025–33). The most interesting discovery of a ballad unknown to Child is perhaps that by F. Sidgwick of *The Bitter Withy* (1905, *NQ*, 10th Series, iv. 84, and *Folk-Lore*, xix. 190–200), studied also by G. H. Gerould (1908, *PMLA* xxiii. 141–67). Other additions to Child's collection may be found in records of folk-song research such as S. Baring-Gould and H. Fleetwood Sheppard, *Songs and Ballads of the West* (1889–91, ed. C. J. Sharp, 1905) and *A Garland of Country Song* (1895); L. E. Broadwood and J. A. Fuller-Maitland, *English County Songs* (1893); L. E. Broadwood, *English Traditional Songs and Carols* (1908); C. J. Sharp and C. L. Marson, *Folk-Songs from Somerset* (1904–9); C. J. Sharp and others, *Folk-Songs of England* (1908–12); A. Williams, *Folk-Songs of the Upper Thames* (1923), and the *Journal of the Folk-Song* (later *Folk Song and Folk Dance*) *Society* (1899 onwards). C. J. Sharp, *English Folk-Song, Some Conclusions* (1907), and F. Kidson and M. Neal, *English Folk-Song and Dance* (1915) are summaries of results.

O. Dame Campbell and C. Sharp, *English Folk Songs from the Southern Appalachians* (New York, 1917), throws light upon English and Scottish conditions as well as upon those of America. American ballads are outside the scope of this chapter. A bibliography of them is in G. H. Gerould, *The Ballad of Tradition* (1932), 297.

General studies of the problems of Balladry are L. Lemcke, 'Die traditionellen schottischen Balladen' (1862, *Jahrbuch für romanische und englische Literatur*, iv); A. Lang (1880, Ward's *English Poets*, i. 203–9), (1901, 1904, *Chambers's Cyclopaedia of English Literature*), (1910, *Encyclopaedia Britannica*); F. J. Child (1893, *Johnson's Cyclopaedia*); F. B. Gummere, *Old English Ballads* (Boston, 1894), *The Ballad and Communal Poetry* (1897, Harvard Studies and Notes in Philology and Literature, v), *The Beginnings of Poetry* (1901), 'Primitive Poetry and the Ballad' (1903–4, *MP* i. 193–202, 217–34, 373–90), *The Popular Ballad* (1907), 'The Ballad' (1908, *Cambridge History of Literature*, ii. 449–74); T. F. Henderson, *Scottish Vernacular Literature* (1898, 1910), *The Ballad in Literature* (1912); E. Flügel, 'Zur Chronologie der englischen Balladen' (1899, *Anglia*, xxi. 312–58); A. Heusler, *Lied und Epos* (Dortmund, 1905); W. M. Hart, 'Professor Child and the Ballad' (1906, *PMLA* xxi. 755–807), *Ballad and Epic* (1907, Harvard Studies); H. Hecht, 'Neuere Literatur zur englisch-schottischen Balladendichtung' (1906, *ES* xxxvi. 370–84); John Meier, *Kunstlieder im Volksmunde* (Halle, 1906); E. Duncan, *The*

Story of Minstrelsy (1907); W. P. Ker, 'On the History of the Ballads, 1100–1500' (1909, *Proc. British Academy*, iv. 179); F. E. Bryant, *A History of English Balladry* (Boston, 1913); P. Barry, 'The Transmission of Folk-Song' (1914, *J. of American Folk-Lore*, xxvii. 67–76); F. Sidgwick, *The Ballad* (1914); A. Beatty, 'Ballad, Tale, and Tradition' (1914, *PMLA* xxix. 473–98); J. C. H. R. Steenstrup, *The Mediaeval Popular Ballad* (Boston, tr. 1914); J. R. Moore, 'The Influence of Transmission on English Ballads' (1916, *MLR* xi. 385–408); L. Pound, *Poetic Origins and the Ballad* (1921, largely reprints of articles in periodicals), 'The Term: "Communal" ' (1924, *PMLA* xxxix. 440–54), 'On the Dating of the English and Scottish Ballads' (1932, *PMLA* xlvii. 10–16); G. H. Gerould, 'The Making of Ballads' (1923, *MP* xxi. 15–28), *The Ballad of Tradition* (1932); L. McWatt, *The Scottish Ballads and Ballad Writing* (Paisley, 1923); A. Keith, 'Scottish Ballads: Their Evidence of Authorship and Origin' (1926, *English Association, Essays and Studies*, xii. 100–19); R. Graves, *The English Ballad* (1927); A. H. Tolman, '*Mary Hamilton*; the Group Authorship of Ballads' (1927, *PMLA* xlii. 422–32); S. B. Hustvedt, *Ballad Books and Ballad Men* (Cambridge, Mass., 1930); G. Humbert, 'Literarische Einflüsse in schottischen Volksballaden' (1932, *Studien zur englischen Philologie*, lxxiv); W. Schmidt, 'Die Entwicklung der englisch-schottischen Volksballaden' (1933, *Anglia*, lvii. 1–77, 113–207, 277–312); F. Panke, *Die schottischen Liebesballaden. Ein Beitrag zur Entstehung von Variantenbildung* (Berlin, 1935); A. Taylor, 'The Themes Common to English and German Balladry' (1940, *MLQ* i. 23–35). Some controversial topics may be studied with advantage in the contributions of Professors Gummere, Pound, and Gerould.

On Robin Hood are J. Ritson, *Robin Hood* (1795, 1832, 1885); J. M. Gutch, *A Lytyll Geste of Robin Hode* (1847); R. Fricke, 'Die Robin-Hood-Balladen' (1883, *Archiv*, lxix. 241–344); L. Fränkel, 'Zur Geschichte von Robin Hood' (1892, *ES* xvii. 316–17); R. Kiessman, *Untersuchungen über die Motive der Robin-Hood Balladen* (Halle, 1895); W. H. Clawson, *The Gest of Robin Hood* (Toronto, 1919); 'Notes on the Name Robin Hood' (1927, 7 Apr. and 9 May, *TLS*). The Robin Hood plays of *c.* 1475 and *c.* 1560 are edited by J. M. Manly (1897, *Specimens of the Pre-Shakesperean Drama*, i. 279–88) and W. W. Greg (1908, *Malone Soc. Collections*, i. 117–36). The relation of Robin Hood to English folk-revels is shown in E. K. Chambers, *The Mediaeval Stage* (1903) and *The English Folk-Play* (1933), and that to Scottish ones in A. J. Mill, *Mediaeval Plays in Scotland* (1927).

Other special topics are considered in K. Nessler, 'Geschichte der Chevy Chase' (1911, *Palaestra*, cxii); A. Saalbach, *Enstehungsgeschichte der schottischen Volksballade, Thomas Rymer* (Halle, 1915); G. R. Stewart, 'The Meter of the Popular Ballad' (1925, *PMLA* xl. 933–62); J. W. Hendren, *A Study of Ballad Rhythm* (1936); W. Christie, *Traditional Ballad Airs* (1876–81); F. Kidson, *Traditional Tunes* (1891); S. B. Hustvedt, *A Melodic Index of Child's Ballad Tunes* (Berkeley, 1936); B. Fehr, *Die formelhaften Elemente in den alten englischen Balladen* (Basle, 1900); G. M. Miller, 'The Dramatic Element in the Popular Ballad' (n. d., *Univ. of Cincinnati Bulletin* 19); P. Schütte, *Die Liebe in den englischen und schottischen Balladen* (Halle, 1906); F. Görbing, 'Beispiele von realisierten Mythen in den englischen und schottischen Balladen' (1901, *Anglia*, xxiii. 1–13); G. Rüdiger, *Zauber und Aberglaube in den englisch-schottischen Volksballaden* (Halle, 1907); K. Ehrke, *Das Geistermotiv in den schottisch-englischen Volksballaden* (Marburg, 1914); H. M. Belden, 'The Relation of Balladry to Folk-Lore' (1911, *Journal of American Folk-Lore*, xxiv); L. C. Wimberly, *Folk-Lore in the English and Scottish Ballads* (Chicago, 1928).

On Continental Balladry, it must be sufficient to cite W. P. Ker, 'On the Danish Ballads' (1904, 1908, *Scottish Historical Review*, i. 357–78; v. 385–401; repr. 1925, in *Collected Essays*), and W. J. Entwistle, *European Balladry* (1939). The outstanding collection of Danish ballads is S. Grundtvig and A. Olrik, *Danmarks gamle Folkeviser* (1853–1920, Copenhagen). Entwistle and Gerould (1932, *supra*) give further bibliography.

CHAPTER IV   MALORY

To the Bibliographies and general works on English Literature, cited on pages 206–7, the following may be added.

The literature on Sir Thomas Malory is extensive. Early editions of the Arthur romance were printed by William Caxton (1485), Jan van Wynkyn de Worde (1498, 1529), William Copland (1557), Thomas East (*c.* 1585), William Stansby for Jacob Bloome (1634). The most important modern ones are those of Robert Southey (1817), Thomas Wright (1856 and later), Edward Strachey (1868 and later), H. O. Sommer(1889–91) with essay by A. Lang, John Rhŷs (1893 and later), I. Gollancz (1897), A. W. Pollard (1900 and later). Selections are by J. W. Hales (1893, *English Prose*, ed. Henry Craik, with general Introduction by W. P. Ker, i. 60–76), A. T. Martin (1897), W. E. Mead (Boston, 1897, 1901), H. Wragg (1912). These are all dependent on Caxton. The discovery of the Winchester MS. is discussed by W. F. Oakeshott in *The Times* (1934, 5 Aug.) and *The Times Literary Supplement* (1934, 27 Sept.), and by E. Vinaver (1935, *John Rylands Library Bulletin*, xix), and an edition, based on it, by Professor Vinaver is in preparation. Before the discovery, he had already written the most comprehensive study of the author and his sources yet available in his *Malory* (1929) which may be supplemented by his *Le Roman de Tristan et Iseut dans l'Œuvre de T. M.* (Paris, 1925). Other studies are in G. Paris, *Merlin* (1886, *SATF*, lxx–lxxii); E. Löseth, *Le Roman en Prose de Tristan, le Roman de Palamède et la compilation de Rusticien de Pise* (Paris, 1890); C. S. Baldwin, *The Inflections and the Syntax of the M.A. of Sir T. M.* (Boston, 1894) and 'The Verb in the M.A.' (1895, *MLN* x. 46–7); E. Wechssler, *Über die verschiedenen Redaktionen des Robert de Boron zugeschriebenen Graal-Lancelot Cyklus* (Halle, 1897); M. Schuler, *Sir T. M.'s M.A. und die englische Arthurdichtung des xix. Jahrhunderts* (Strassburg, 1900); A. C. L. Brown, 'Balin and the Dolorous Stroke' (1909, *MP* vii. 203–6); W. H. Schofield, *Chivalry in English Literature* (1912); H. O. Sommer, 'Die Abenteuer Gawains, Ywains, und Le Morholts mit den drei Jungfrauen' (1913, *Beihefte zur Zeitschrift für romanische Philologie*, xlvii, p. xxvii); V. D. Scudder, *Le M.A. of Sir T. M.* (1917, 1921); E. Vettermann, 'Die Balen Dichtungen und ihre Quellen' (1918, *Beihefte zur Zeitschrift für romanische Philologie*, lx. 52–84); E. K. Chambers, *Sir T. M.* (1922, English Association, repr. (1933) with note in *Sir Thomas Wyatt and Some Collected Studies*, 21–45); L. A. Hibbard, 'M's Book of Balin' (Paris and New York, 1927, *Mediaeval Studies in Memory of G. S. Loomis*, 175–95); M. D. Fox, 'Sir T. M. and the Piteous History of the M.A.' (1928, *Arthuriana*, i . 30–6); N. S. Aurner, 'Sir Thomas Malory—Historian?' (1933, *PMLA* xlviii. 362–91); E. Whitehead, 'On Certain Episodes in the Fourth Book of M's M.A.' (1933, *Medium Aevum*, ii. 199); G. R. Stewart, 'English Geography in M's M.A.' (1935, *MLR* xxx. 204–9); R. S. Loomis, 'M's Beaumains' (1939, *PMLA* liv. 656–68);

C. F. Bühler, 'Two Caxton Problems' (1939, *Library*, xx. 266–71); R. H. Wilson, 'The Fair Unknown in Malory' (1943, *PMLA* lviii. 1–21). The relation of Malory's work to the alliterative *M.A.* poem is discussed by T. Vorontzeff, 'M's Story of Arthur's Roman Campaign' (1937, *Medium Aevum*, vi. 99); E. Vinaver and E. V. Gordon, 'New Light on the Text of the Alliterative M.A.' (1937, *Medium Aevum*, vi. 111); I. D. O. Arnold, 'M's Story of A's Roman Campaign' (1938, *Medium Aevum*, vii. 74); and that to the stanzaic *M.A.* by J. D. Bruce, 'The M.E. Metrical Romance Le M.A.' (1901, *Anglia*, xxiii. 67–100), and in his edition of the poem (1903, EETS e.s. lxxxviii), by H. O. Sommer (1906, *Anglia*, xxix. 529–38), and again by J. D. Bruce 'A Reply to Dr. Sommer' (1907, *Anglia*, xxx. 209–16). On Sir Thomas Malory of Warwickshire are G. L. Kittredge, 'Sir T. M.' (1894, *Johnson's Universal Cyclopaedia*), 'Who was Sir T. M.?' (1897, *Studies and Notes in Philology and Literature*, v. 85–106) and *Sir T. M.* (1925, Barnstable, privately printed); T. W. Williams, 'Sir T.M.' (1896, *Athenaeum*, 64–5, 98–9) and *Sir T. M. and Le M.A.* (1909, Bristol, privately printed); A. T. Martin, 'Sir T. M.' (1897, *Athenaeum*, 353–4) and 'The Identity of the Author of the M.A.' (1898, *Archaeologia*, lvi. 165–77); E. Hicks, *Sir T. M., his Turbulent Career* (Cambridge, Mass., 1928); A. C. Baugh, 'Documenting Sir T. M.' (1933, *Speculum*, viii. 3–29).

INDEX